D1730376

Developing Tsunami-Resilient Communities

We dedicate this book to the victims of tsunamis, and especially those affected by the catastrophic tsunami of December 26, 2004

Developing Tsunami-Resilient Communities

The National Tsunami Hazard Mitigation Program

Edited by
E. N. BERNARD

Reprinted from *Natural Hazards,* Volume 35(1), 2005

A C.I.P. catalogue record for this book is available from the Library of Congress

ISBN 1-4020-3353-2

Published by Springer,
P.O. Box 17, 3300 AA Dordrecht, The Netherlands

Sold and distributed in North, Central and South America
by Springer,
101 Philip Drive, Norwell, MA 02061, USA

In all other countries, sold and distributed
by Springer,
P.O. Box 322, 3300 AH Dordrecht, The Netherlands

Printed on acid-free paper

Printed in the Netherlands

Table of contents

Natural Hazards (2005) 35: 1–4

Preface

As the world grieves over the catastrophic loss of humanity from the 26 December 2004 tsunami, we must resolve to learn from nature's lessons. This issue provides a framework and a set of tools to develop communities that are resilient to tsunami. This collection of papers represents a starting point on our new journey toward a safer world.

The history of tsunami hazard mitigation tracks well with the history of destructive tsunamis in the United States. Following the 1946 Alaska generated tsunami that killed 173 people in Hawaii, the Pacific Tsunami Warning Center was established in Hawaii by a predecessor agency to the National Oceanic and Atmospheric Administration (NOAA). Following the 1960 Chilean tsunami that killed 1,000 people in Chile, 61 in Hawaii, and 199 in Japan, the United States formed the Joint Tsunami Research Effort (JTRE) and staffed the International Tsunami Information Center (ITIC) in Hawaii. JTRE was formed to conduct research on tsunamis while ITIC, sponsored by the United Nations, was formed to coordinate tsunami warning efforts of the Pacific Countries. Many research and mitigation efforts were focused on the distant tsunami problem. Following the 1964 Alaskan tsunami that killed 117 in Alaska, 11 in California, and 4 in Oregon, the U.S. was confronted with the local tsunami problem. In response, the U.S. established the Alaska Tsunami Warning Center in Palmer, Alaska.

In 1992, a Ms 7.2 earthquake in California generated a tsunami that killed no one. However, the earthquake was the first subduction zone earthquake recorded on the U.S. west coast with modern instruments. The earthquake triggered concern that larger earthquakes could generate large local tsunamis along the heavily populated west coast. In response to the local tsunami threat, the National Tsunami Hazard Mitigation Program (NTHMP) was formed in 1997 and the Alaska Tsunami Warning Center was renamed the West Coast and Alaska Tsunami Warning Center.

In 1994, Congress asked NOAA, responsible for issuing tsunami warnings to the U.S., to assess tsunami awareness and preparedness of the west coast for local tsunamis. NOAA held three workshops, which led to the publication of technical reports with recommendations for improvements. Congress asked NOAA to lead a group of representatives from the U.S. Geological

Survey (USGS), the Federal Emergency Management Agency (FEMA), and from emergency management agencies in the states of Alaska, California, Hawaii, Oregon, and Washington to review these recommendations and formulate a plan of action. The group formed a State/Federal partnership, the National Tsunami Hazard Mitigation Program, which developed a 5-year implementation plan, including a budget. Congress, led by Senator Hatfield of Oregon, Senator Inouye of Hawaii, and Senator Stevens of Alaska, funded the implementation plan beginning in 1997. Because of Senator Hatfield's retirement, Senators Inouye and Stevens have continued to champion the effort. The collection of 12 papers presented in this issue document the accomplishments of the NTHMP, which is now a permanent part of NOAA operations. The Program is viewed as a model for State/Federal Partnerships.

This issue is organized into four sections: (1) a summary paper that gives an overview of the NTHMP with reference to papers in the issue for further discussions, (2) four papers on warning guidance that detail the development of a tsunami forecasting capability and upgrades in tsunami warning operations, (3) two papers on hazard assessment that describe the development of tsunami inundation maps and their use in tsunami warnings, and (4) five papers on mitigation that describe the concept of tsunami-resilient communities. Two innovations of the program are to create a tsunami forecasting capability and to introduce the concept of tsunami-resilient communities. Combined, these innovations constitute a major advance in tsunami hazard mitigation for both local and distant tsunamis.

The reader should understand that many of the accomplishments reported have their origins with scientists at the Joint Tsunami Research Effort. For example, the capability of forecasting tsunamis using deep ocean tsunami measurements complemented by numerical modeling originated from the research of Gaylord Miller, Martin Vitousek, Harold Loomis, and Robert Harvey. It took about 30 years to transform the idea of measuring tsunamis in the deep ocean to actually reporting such data in real time. The technical feat of transmitting data from an instrument on the sea floor to a tsunami warning center in real time required exceptionally creative engineering that is carefully documented in the González *et al.* paper. The new tsunami measuring technology has given science a new instrument – the tsunameter – that provides tsunami researchers and practitioners with the basic information to understand and predict tsunamis. In 2003, a real-time tsunameter detected a non-destructive tsunami which led to the early cancellation of a tsunami warning and averted an unnecessary evacuation in Hawaii. For this significant feat, the Department of Commerce awarded NOAA its highest award, the Gold Medal. The second technology required to predict tsunamis is numerical models of tsunami dynamics. The paper by Titov *et al.* describes the use of tsunameter data as input for numerical models to forecast tsunamis. The two papers document an amazing technological development to create a tsunami forecasting capability for the United States. More importantly, the tsunameter/model combination has transformed the warning function from tsunami detection to tsunami forecasting. In operational use, the tsunameter/model will eventually lead to accurate tsunami forecasts that save lives. Accurate forecasts will also lead to fewer false alarms that cost in lost productivity and in lost confidence in the warning system.

The ability to identify tsunami hazard zones provides at-risk coastal communities with the most basic tool for tsunami preparedness. Once a community has the tsunami hazard zone identified, evacuation maps can be developed enabling residents to safely and efficiently escape tsunami dangers. The seminal paper by González *et al.* reviews the procedure developed to produce tsunami inundation maps and provides a set of best practices for ensuring the scientific quality of the maps.

The concept of tsunami-resilient communities was created to provide direction and coordination for tsunami mitigation activities in the absence of a disaster. Early recognition that no mitigation effort would succeed without involvement and support of local communities provided the impetus to start a dialog between state and local emergency managers/planners/responders and other local decision makers. The paper by Jonientz-Trisler *et al.* gives a detailed account of the activities of the five states in establishing the dialog and subsequent actions. Specific tools that have been developed include education (Dengler), community design (Eisner), and tsunami warning alerts (Crawford). The paper by Johnson *et al.* assesses the effectiveness of these mitigation activities. Combined, these papers document a mitigation effort that recognizes that the ultimate responsibility for sustained mitigation is with the users of the coastal environment. With proper attention to coastal land use and proper response to tsunami warnings, coastal communities can survive the next local or distant tsunami.

EDDIE BERNARD
Editor

Acknowledgements

The editor thanks the authors for their scientific contributions with a focus on society. I feel fortunate to be included among this distinguished group of writers. The editor expresses warm appreciation for the efforts of reviewers including: Linda Anderson-Berry, Patricia Bolton, Ted Buehner, Darrell Ernst, Eric Geist, James Godfrey, Marjorie Greene, Bruce Jaffe, Gail Kelly, Laura Kong, Philip Liu, Douglas Luther, Richard McCarthy, Mike Mahoney, Hugh Milburn, Sarah K. Nathe, Emile Okal, Kenji Satake, Costas

Synolakis, Paul Whitmore, Tyree Wilde, and Al Yelvington. Their community service has greatly improved the quality of the papers. Ann Thomason's assistance in obtaining and tracking the reviews is gratefully acknowledged. Most importantly, Ryan Layne Whitney's editorial skill and attention to detail in the final editing of the issue is the main factor in the quality of the final publication. Ryan has provided priceless counsel on many technical details along each step of the publication process. He has been an incredible travel partner whom I greatly appreciate and admire.

Natural Hazards (2005) 35: 5–24

The U.S. National Tsunami Hazard Mitigation Program: A Successful State–Federal Partnership

EDDIE N. BERNARD
NOAA/Pacific Marine Environmental Laboratory, Seattle, Washington 98115, USA, and first Chair, NTHMP Steering Group (Tel: +1-206-526-6800; Fax: +1-206-526-4576; E-mail: eddie.n.bernard@noaa.gov)

(Received: 1 October 2003; accepted: 19 April 2004)

Abstract. The U.S. National Tsunami Hazard Mitigation Program (NTHMP) is a State/Federal partnership created to reduce tsunami hazards along U.S. coastlines. Established in 1996, NTHMP coordinates the efforts of five Pacific States: Alaska, California, Hawaii, Oregon, and Washington with the three Federal agencies responsible for tsunami hazard mitigation: the National Oceanic and Atmospheric Administration (NOAA), the Federal Emergency Management Agency (FEMA), and the U.S. Geological Survey (USGS). In the 7 years of the program it has,

1. established a tsunami forecasting capability for the two tsunami warning centers through the combined use of deep ocean tsunami data and numerical models;
2. upgraded the seismic network enabling the tsunami warning centers to locate and size earthquakes faster and more accurately;
3. produced 22 tsunami inundation maps covering 113 coastal communities with a population at risk of over a million people;
4. initiated a program to develop tsunami-resilient communities through awareness, education, warning dissemination, mitigation incentives, coastal planning, and construction guidelines;
5. conducted surveys that indicate a positive impact of the program's activities in raising tsunami awareness.

A 17-member Steering Group consisting of representatives from the five Pacific States, NOAA, FEMA, USGS, and the National Science Foundation (NSF) guides NTHMP. The success of the program has been the result of a personal commitment by steering group members that has leveraged the total Federal funding by contributions from the States and Federal Agencies at a ratio of over six matching dollars to every NTHMP dollar. Twice yearly meetings of the steering group promote communication between scientists and emergency managers, and among the State and Federal agencies. From its initiation NTHMP has been based on the needs of coastal communities and emergency managers and has been results driven because of the cycle of year-to-year funding for the first 5 years. A major impact of the program occurred on 17 November 2003, when an Alaskan tsunami warning was canceled because real-time, deep ocean tsunami data indicated the tsunami would be non-damaging. Canceling this warning averted an evacuation in Hawaii, avoiding a loss in productivity valued at $68M.

Key words: tsunami hazard reduction, tsunami, tsunameters, State-Federal partnerships

Abbreviations: CSZ – Cascadia subduction zone, FEMA – Federal Emergency Management Agency, FRP – Federal Response Plan, NEES – Network for Earthquake Engineering

Simulation, NOAA – National Oceanographic and Atmospheric Administration, NSF – National Science Foundation, NTHMP – National Tsunami Hazard Mitigation Program, NWR – NOAA Weather Radio, PTWC – Pacific Tsunami Warning Center, TIME – Tsunami Inundation Mapping Effort, USGS – U.S. Geological Survey, WC/ATWC – West Coast/ Alaska Tsunami Warning Center

1. Background

On 25 April 1992, a magnitude 7.1 (Mw) earthquake occurred on California's North Coast, producing a small tsunami that was detected along the California and Oregon coast and in Hawaii (González and Bernard, 1993). The location and orientation of rupture confirmed the possibility that the Cascadia subduction zone (CSZ) could produce strong earthquakes and local tsunamis. This earthquake/tsunami potential raised the concerns of State and Federal agencies responsible for disaster planning and response. On 4 October 1994, a Pacific-wide tsunami warning was issued after a major earthquake in the Kuril Islands produced a tsunami. The warning caused confusion among emergency managers, uneven response of coastal communities, and costly evacuations that resulted in angry recriminations when no significant tsunami arrived. The 1994 tsunami emphasized the need to reduce false warnings and improve communication and coordination between states and the warning centers.

These two recent tsunamis raised concerns about U.S. tsunami preparedness. As a result, the Senate Appropriations Committee directed NOAA, the federal agency responsible for issuing tsunami warnings, to formulate a plan for reducing the tsunami risks to coastal residents. Within 10 months, NOAA hosted three tsunami workshops on hazard assessment (Bernard and González, 1994), warnings (Blackford and Kanamori, 1995), and education (Good, 1995) involving over 50 scientists, emergency planners, and emergency operators from all levels of governments and universities during which 12 recommendations were formulated for improvements. These recommendations were submitted to the Senate Committee in March 1995. By October 1995 the U.S. Senate Appropriations Committee directed NOAA to form and lead a Federal/State Working Group to develop an action plan with budget to address tsunami mitigation based on the 12 recommendations. The first meeting of the Tsunami Hazard Mitigation Federal/State Working Group, with representatives from the States of Alaska, California, Hawaii, Oregon, and Washington and three Federal agencies – NOAA, FEMA, and USGS – was held in February 1996 (Bernard, 1998). The Tsunami Hazard Mitigation Implementation Plan developed by the Group addressed four primary issues of concern to the states.

- Quickly confirm potentially destructive tsunamis and reduce false alarms.
- Address local tsunami mitigation and the needs of coastal residents.
- Improve coordination and exchange of information to better utilize existing resources.
- Sustain support at state and local level for long-term tsunami hazard mitigation.

From these issues, the Plan established three fundamental areas of effort and five specific Program elements at a cost of \$2.3M/year (Tsunami Hazard Mitigation Federal/State Working Group, 1996):

3.1. Tsunami Hazard Assessment
  1. Produce Inundation Maps

3.2. Tsunami Warning Guidance
  2. Improve Seismic Networks
  3. Deploy Tsunami Detection Buoys
  4. Improve State/NOAA Coordination and Technical Support for Tsunami Warnings

3.3. Mitigation
  5. Develop State/Local Tsunami Hazard Mitigation Programs

2. Governance

The Plan established a 17-member steering group, with representatives from the five Pacific States of Alaska, California, Hawaii, Oregon, and Washington, NOAA, FEMA and the USGS, to oversee the NTHMP (see Figure 1). Two individuals, one from the State emergency management agency and another from the State geotechnical agency (the State counterpart to the USGS) represent each state. Similarly, at least two representatives for each Federal Agency are Steering Group members. Steering group members prepare and defend proposals to implement the three elements identified in the implementation plan. The Steering Group meets twice a year to report progress on the five elements and to present proposals for future activities. An eight-member Executive Committee is responsible for funding decisions, with one vote allotted to each of the five States and three Federal Agencies and, when required, a ninth tie-breaking vote allocated to the NTHMP Chairperson. The Chair is elected by all members. Through this proposal, review, and adjustment process, modifications are continuously made to meet NTHMP goals. This results-driven process has led to the rapid development of numerous mitigation tools and products, described in following sections, that are expected to reduce the impact of future tsunamis on U.S. coastlines.

Funding of \$2.3M was provided on a fiscal year-by-year basis from 1998–2001 through the Congressional add-on process. Federal dollars were

Figure 1. Logos of partners.

matched with $6 state/agency in-kind or dollar contribution for every program dollar (Bernard, 2001). In 2001, a team of five external experts was tasked to conduct a 1-day review of the accomplishments of the first 5 years of NTHMP. The review team was favorably impressed with the quality and relevance of the accomplishments and recommended increased funding. Based on reviewer comments and evolving state needs, NTHMP group members formulated goals for the 5 years beginning with 2002 (see Future section in this paper). Because of the valuable contributions made by the external reviewers, they were asked to serve as members of a NTHMP technical advisory committee. In 2001, NTHMP became a part of the NOAA base budget and in 2002, the budget was increased to $4.3 million. By the end of 2003, the program had received a total of $18.1 million. The NTHMP web site (http://www.pmel.noaa.gov/tsunami-hazard/) contains the minutes for Steering Group meetings, steering group members, proposals, and progress reports for each element as well as links to other tsunami sites of interest.

3. Accomplishments

3.1. TSUNAMI HAZARD ASSESSMENTS

From the beginning of the NTHMP there has been unanimous agreement among Steering Group members that inundation and evacuation maps are the fundamental basis of local tsunami hazard planning. The initial goals of NTHMP were to provide all at-risk U.S. coastal communities with a preliminary inundation map within 3 years using 1-D mapping algorithms developed by the University of Hawaii and having local city and county engineers run the models. There were two significant difficulties. Local city and county engineers were not available to produce the maps, and the 1-D models were considered inadequate by the states to accurately map coastal communities often located in harbors and/or near estuaries. The original plan was revised to provide inundation maps for the highest-priority communities using 2-D modeling technology. Twenty-two inundation modeling and mapping efforts have been completed for 113 communities with an estimated population at risk of 1.2 million people (González *et al.*, this issue). Evacuation maps based on the inundation maps have been completed for 23 areas (González *et al.*, this issue). Figure 2 is an example of an evacuation map for Rockaway Beach, Oregon based on an inundation map. Other examples of tsunami inundation activities can be found in Borrero *et al.* (2003).

The Tsunami Inundation Mapping Effort (TIME) Center was created to develop an infrastructure to support tsunami inundation modeling. TIME staff assists states and modelers in acquiring bathymetric data and assessing its quality, with the setup and execution of grid/model software and with special processing and analysis of model output to provide useful and consistent final products. A procedural checklist has emerged as the result of experience gained from the production of inundation maps for five states over the past 7 years (González *et al.*, this issue). The checklist is.

1. Prioritize a list of communities to map.
2. Identify specific potential tsunami sources.
3. Develop a computational grid having a resolution less than 50 m for the selected community.
4. Draft a project time line.
5. Complete the map.
6. Publish the map.

Publishing each map documents the procedures used for production and establishes a technical and physical baseline from which future adjustments will be made. To date, 31 documents have been published to record the map production methodology.

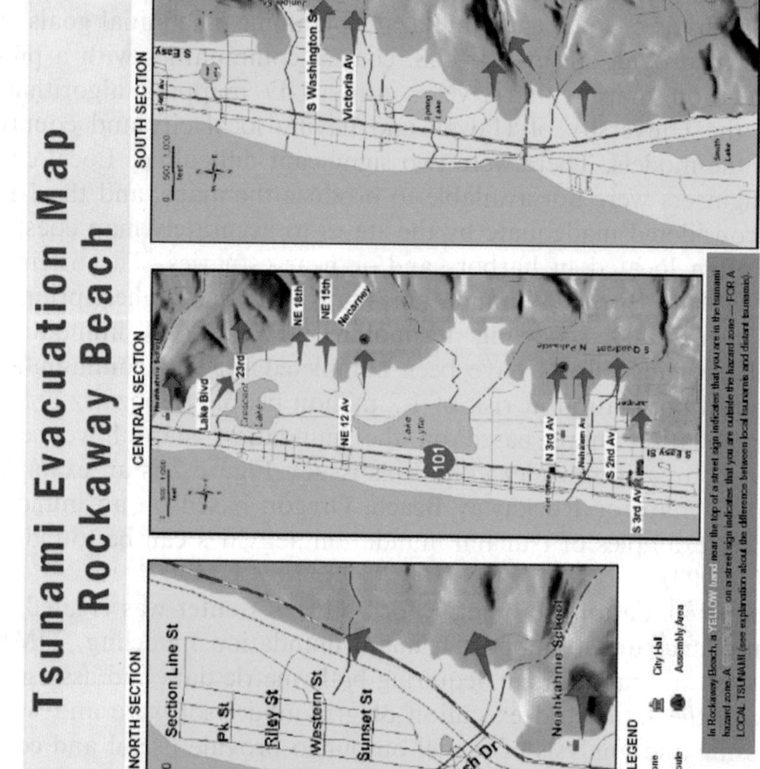

Figure 2. Rockaway Beach Evacuation map.

Figure 3. World's largest tsunami wave basin facility.

A direct result of tsunami inundation mapping is that community planners are now investigating what activities are appropriate in these threatened coastal areas. Serendipitously, the new NSF Network for Earthquake Engineering Simulation (NEES) program has funded the construction of a tsunami wave basin facility at Oregon State University. The facility (see Figure 3) became available for research in September, 2003 (http://www.nees.org/). Also, the NEES grand challenge report from the National Research Council establishes a coordinated effort between NTHMP and NEES (National Research Council, 2003). The NTHMP/NEES partnership has placed priority on the use of the new tsunami facility for determination of forces on structures for construction guidance. This example illustrates the value of a long-term state/federal partnership that can coordinate opportunities as new federal or state programs develop.

3.2. TSUNAMI WARNING GUIDANCE

NOAA operates two tsunami warning centers: the Richard H. Hagemeyer Pacific Tsunami Warning Center (PTWC) in Hawaii and the West Coast/Alaska Tsunami Warning Center (WC/ATWC) in Alaska. Each center monitors seismic data for the Pacific basin and issues alert bulletins (warning, watch, advisory) if an earthquake's magnitude exceeds a threshold value. Sea level data from a coastal tide gage network is used to continue, upgrade, or cancel the warnings and watches. Alert messages are transmitted to the states via a variety of dissemination methods. Three elements of NTHMP address

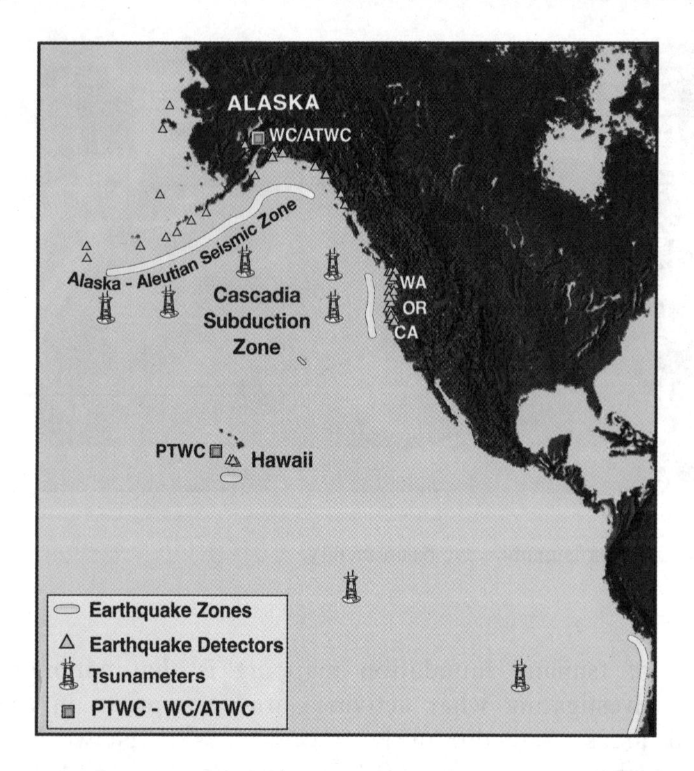

Figure 4. Map indicating locations of tsunameters and seismometers.

improving the speed, quality, and dissemination of alert messages, and providing education essential for appropriate response.

3.2.1. *Improve Seismic Networks*

The NTHMP called for (a) installation of real-time broadband seismic stations to improve seismic coverage, (b) telemetry upgrades to the warning centers and integration of existing regional networks to provide faster and more robust seismic data, and (c) shortening of information dissemination time to emergency services agencies.

The USGS developed the seismic project to meet the NTHMP goals (Oppenheimer *et al.*, this issue). In the first 7 years of the project, 53 broadband seismic stations (see Figure 4 for locations) were installed with specialized software that allowed the warning centers to access additional seismic data from nine other networks supported by the USGS. The seismic project also strengthened the telecommunications links between the networks and designed a level of redundancy into the system to provide reliable backup in case of power or communications failures.

The tsunami warning centers now have real-time, high-dynamic range, broadband seismic data from regions of the U.S. where tsunamigenic earthquakes can occur, as well as from seismic stations around the world. The average response time of the WC/ATWC to issue messages for the period 1982–1998 was 10.6 minutes for U.S. earthquakes. The 28 February 2001, Mw 6.8 Seattle earthquake in Washington state provided an opportunity to evaluate the response of the upgraded system. Because of the NTHMP seismic stations installed in the Pacific Northwest, WC/ATWC was able to issue a message in 2 minutes. The additional seismic data has also significantly reduced the time to issue warnings for teleseismic events. Formerly it took up to 90 minutes to locate and determine the size of earthquakes in the western Pacific basin; it now takes a maximum of 25 minutes (McCreery, this issue).

3.2.2. *Deploy Tsunami Detection Buoys*

Coastal tide gages are useful for detecting the presence of tsunamis, but are unreliable for predicting tsunami wave heights at other locations. Tsunamis are altered by nearshore bathymetry and coastal topography and do not provide a direct measurement of tsunami energy. While earthquake location and magnitude is the quickest determinant of a potentially tsunamigenic earthquake, magnitude is a poor predictor of potentially damaging waves. NOAA's Tsunameter Project is an effort of the NTHMP to quickly confirm or cancel a tsunami warning (González *et al.*, this issue). This is of particular concern to the State of Hawaii, where Hawaii Civil Defense must make evacuation decisions within an hour of a large earthquake in the Alaska–Aleutian subduction zone. False evacuations are costly both in terms of dollar expenses and credibility.

The Tsunameter Project has installed pressure gage sensors on the deep ocean sea floor that can measure and detect a tsunami of only a few millimeters amplitude in the open ocean (González *et al.*, this issue). Tsunameters have been installed at six sites – three offshore of the Alaska–Aleutian trench, two offshore of Washington and Oregon, and one in the equatorial Pacific far offshore of Ecuador (see Figure 4 for locations). The Chilean Government has purchased one tsunameter from NOAA and has deployed it off their coastline at 20°S. The Chilean tsunameter increased the size of the array from six to seven tsunameters (see Figure 4 for locations). Each instrument package consists of a sea floor pressure sensor in acoustic contact with an anchored buoy that transmits the ocean bottom data to a GOES satellite (see Figure 5 for details) where the data are disseminated to the warning centers and to the internet (http://www.ndbc.noaa.gov/dart.shtml). Data from tsunameters, free of the coastal effects, provide accurate forecasts of tsunamis by assimilating real-time tsunameter data into nested numerical models.

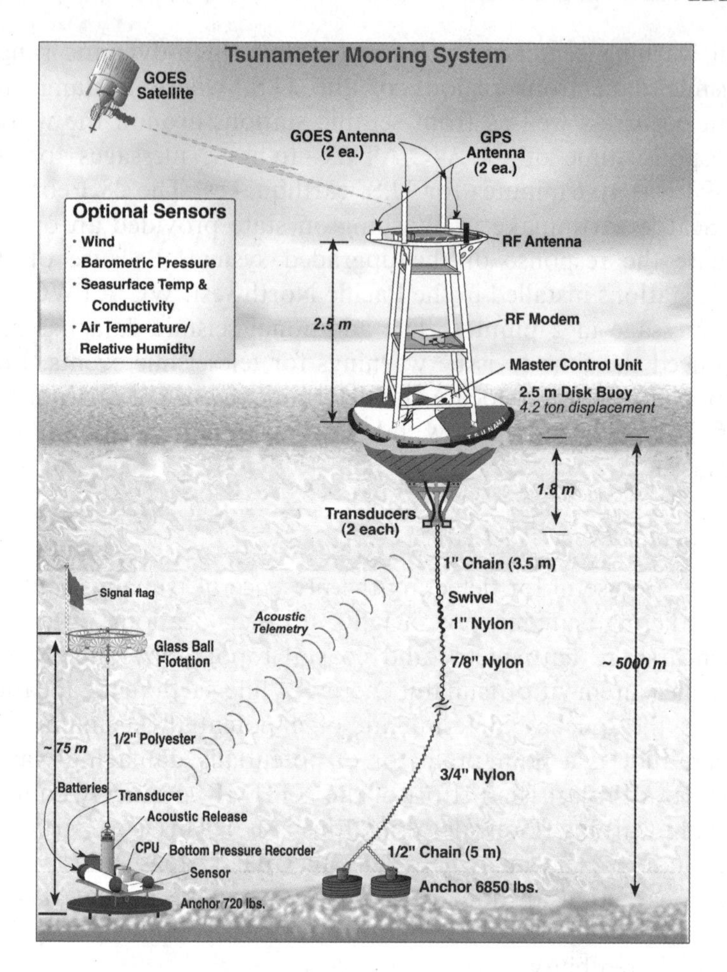

Figure 5. Tsunameter system.

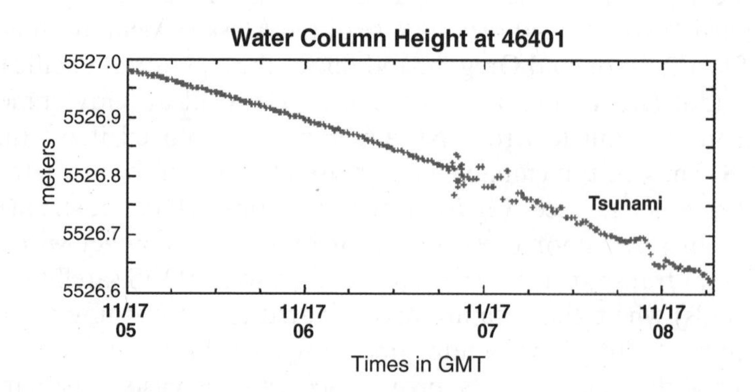

Figure 6. Real-time detection of 17 November 2003 tsunami.

Titov *et al.* (this issue), describes the data/model assimilation methodology which was used for an experimental forecast of the 17 November 2003 Aleutian Island tsunami at Hilo, Hawaii.

McCreery (this issue) reports that the accuracy of earthquake magnitudes have improved and the time to determine location and magnitudes have been reduced due to the improved data from the seismic network. He also reports a reduction in false alarms due to the tsunameter array, two original goals of the NTHMP. One specific example occurred on 17 November 2003 at 0643 UTC, when a Ms 7.5 earthquake in the Aleutian Islands generated a tsunami and a tsunami warning. By 0805, the tsunameter located closest to the earthquake had detected a 2 cm amplitude tsunami (see Figure 6) and sent these data to the NOAA tsunami warning centers and the internet in real time. Based on the tsunameter data and coastal tide data, the warning was canceled at 0812 UTC or 89 minutes after the earthquake origin time.

Quickly canceling tsunami warnings saves millions of dollars in unnecessary evacuations. The State of Hawaii has a policy to evacuate if a tsunami warning is in effect when the tsunami arrival time is within 3.5 hours of the closest Hawaii coastline. The policy was formulated to allow ample time to evacuate Hawaii's busy coastline. For the November 17 tsunami warning, early cancelation avoided an evacuation. The State of Hawaii has estimated that a false alarm evacuation can cause an economic loss of $58.2M (Hawaii Department of Business, Economic Development, and Tourism, 1996). Adjusting for inflation, the avoidance of one false alarm in 2003 would save about $68M. Thus, the $18.1M NTHMP investments in tsunami hazard mitigation have financially benefitted the U.S. More importantly, the upgrades will provide faster, more accurate tsunami forecasts that will ultimately translate into lives saved.

3.2.3. *Improve State/NOAA Coordination and Technical Support for Tsunami Warnings*

The 1994 Kuril Islands tsunami warning illustrated a lack of understanding about the tsunami warning system at the local level, confusion over the format of alert bulletins and frustration with the level of information during a tsunami warning event. To address these concerns, states identified tsunami experts and/or state-wide tsunami coordinators. NOAA provided technical support for state and local emergency managers by encouraging a more active role of the Warning Coordination Meteorologists in the regional National Weather Service Offices. The format of tsunami alert bulletins was changed for clarity and NOAA Weather Radio (NWR) coverage has been improved and has become a principal means of disseminating tsunami information to coastal residents. A recent survey of emergency managers shows significant improvement in understanding the warning system messages. Other activities, including a complete inventory of local tsunami warning systems (Darienzo *et al.*, this issue) and a novel use of NWR for alerting people on beaches (Crawford, this issue) have made NOAA tsunami warnings more effective.

3.3. MITIGATION

Mitigation activities of the NTHMP are coordinated by the Mitigation Subcommittee consisting of the two representatives from each state – an emergency manager and a geoscientist – and chaired by the FEMA program representative. During the first year of NTHMP, a strategic mitigation plan was developed to formulate mitigation strategies and set priorities for funding projects (Dengler, 1998). The strategy recognizes the different tsunami exposure and unique demographic situations of the five Pacific states and the need to incorporate tsunami efforts into existing earthquake and all-hazard mitigation programs. It also emphasizes the variety of mitigation products and projects of differing scope and scale, the need to couple product development with a well-defined method of distribution and dissemination, and the importance of having the support of local populations, without which mitigation success would be compromised. This strategic goal is to develop "tsunami resilient" communities having the following characteristics:

1. *Understands the nature of the tsunami hazard.* Knows the risk that tsunamis, from both near and far sources, pose to its coastal areas.
2. *Has the tools it needs to mitigate the tsunami risk.* Has defined needed mitigation products, has access, and knows how to use them.
3. *Disseminates information about the tsunami hazard.* Has identified vulnerable populations, has materials which include areas at risk and safety, evacuation routes, appropriate response, and has developed a dissemination plan to provide information to all users of the coastal area.
4. *Exchanges information with other at-risk areas.* Supports mitigation efforts through the free exchange of information, products, and ideas with other at-risk areas and learns from the mitigation efforts for other natural hazards.
5. *Institutionalizes planning for a tsunami disaster.* Has incorporated tsunami hazard mitigation elements into their long-term all-hazard management plans and has developed a structure to develop and maintain the support of local populations and decision makers for mitigation efforts.

In the first 5 years of the program, a total of $250,000 was provided to each state to develop and support state tsunami programs. NTHMP funds were matched by over 2 to 1 by the states in dollars and in-kind efforts. The state programs developed individual strategies based on their unique populations, legislative systems, and existing technical framework. All states

Figure 7. Rockaway Beach evacuation signs.

strongly emphasized the input of coastal communities in the development of their programs. A portion of mitigation funding each year was designated for multi-state projects that address the needs of two or more states. Accomplishments of the mitigation program include: adoption of standardized tsunami evacuation and hazard zone signage for all five states (see Figure 7); the TsuInfo Alert newsletter (http://www.wa.gov/dnr/htdocs/ger/tsuinfo/) published six times/year; regional, state, and multi-state workshops; and certification of five communities as TsunamiReady (http://www.wcatwc.gov/tsunamiready/tready.htm).

Jonientz-Trisler *et al.* (this issue) summarizes mitigation activities in their Table I. Of particular interest in the paper are the surveys of emergency managers in 1994, 2001, and 2003 following the issuance of tsunami warnings and subsequent cancelations. The repeat survey in 2001 indicated the level of understanding had doubled; that was attributed to the efforts of the NTHMP. Also of interest is a 2001 survey of the residents' and visitors' perception of tsunami hazards along Washington's coastline (Johnston *et al.*, this issue) that indicates that awareness is high, but levels of preparedness are low. Such surveys are critical to measuring the effectiveness of the NTHMP and defining areas for improvement. Dengler (this issue) discusses the role of education and the media as a key mitigation tool. Her case study of northern

California over 9 years demonstrates the need for educational tools and annual surveys of the effectiveness of these tools to increase tsunami hazard awareness. Eisner (this issue) identifies seven principles recommended in designing a community to become tsunami resilient.

3.4. SUMMARY

Significant progress has been made on the issues of primary concern to the states:

- Quickly confirm potentially destructive tsunamis and reduce false alarms.

Although a tsunami destructive to U.S. shorelines has not occurred since the program's inception, the major upgrades of the tsunami warning system with real-time tsunameter data and seismic data have made significant progress in quickly assessing the danger of an approaching tsunami. At least one false alarm has been averted on 17 November 2003, using the new tsunameter data available to the NOAA tsunami warning centers. The tsunameter data also provide a real-time forecasting capability that differentiates between destructive and non-destructive tsunamis.

- Address local tsunami mitigation and the needs of coastal residents.

The creation of tsunami inundation maps marks a major advance in identifying the portions of coastline susceptible to tsunami flooding. Evacuation maps are then derived from these maps to address local issues of coastal residents. To date, about 30% of the at-risk communities have been mapped. The highest priority for the program should be the completion of the mapping effort.

- Improve coordination and exchange of information to better utilize existing resources.

The impressive leverage of the program of attracting $6 for every $1 of NTHMP funds is a quantifiable improvement in the utilization of existing resources. The primary contribution to this leverage is the seismic network upgrades which makes seismic data available to the NOAA tsunami warning system from all over the world in real time. In addition, standard earthquake products utilizing these data can be easily transferred from the seismology research and development community into tsunami warning operations. Tsunameters now offer tsunami data which will lead to real-time tsunami forecasts.

In 2003, NSF and Oregon State University dedicated the world's largest tsunami wave basin facility to conduct research on tsunami forces on structures. This research effort is coordinated with NTHMP mapping efforts, maximizing the use of existing resources.

– Sustain support at state and local level for long-term tsunami hazard mitigation.

The NTHMP began as a Congressional add-on program. By 2001, program success and progress convinced NOAA and the U.S. Congress that the program should become a permanent part of NOAA's operations. The permanence of the NTHMP will allow a sustained effort of the activities described in the mitigation section of this paper to develop "tsunami resilient" communities. Certification programs such as Tsunami Ready, educational programs, and principles to design tsunami resilient communities along with surveys to measure effectiveness in increasing awareness and preparedness will address the long-term needs of local communities.

4. Future Goals

The NTHMP has achieved the initial goals set forth at inception in 1996. Inundation maps have been published, a tsunameter network and a tsunami forecast capability are in place, and education and outreach have been provided to at-risk coastal populations. In an effort to establish future goals, five experts were asked to review the progress of the NTHMP at the 5-year mark to offer suggestions for improvements. The reviewers included: Professor Hiroo Kanamori, a world renowned seismologist who specializes in tsunami generation, Professor Philip L. F. Liu, an ocean engineer who specializes in numerical modeling of tsunami inundation, Richard J. McCarthy, a geologist who specializes in earthquake/tsunami mitigation with experience in state/federal partnerships, Professor Douglas Luther, a physical oceanographer who specializes in deep ocean wave measurements, and Professor Dennis S. Mileti, a nationally recognized sociologist who specializes in mitigation for natural hazards.

The criteria for evaluation included: Has the program successfully met the goals of the implementation plan? Are the products technically sound? Is the state/federal partnership working? Do you expect the products to have a positive impact on tsunami mitigation? Are plans for the future appropriate?

Overall comments from the reviewers included:

> "It is such a relief to see how much can be accomplished with so little funding when the proper leadership is exercised. NOAA provided that leadership and the program is a model success story on how to form partnerships between the federal, state, and local governments to reduce risk cost effectively."

"NTHMP is an outstanding program." "The program is a model of how academics and state and federal personnel can work together to the great benefit of the U.S. public."

"Two factors stand out as major contributors to the success of the program: strong but inclusive leadership and enthusiastic cooperation among the personnel of the three federal agencies, five state governments, and four academic institutions."

"The progress during the first 5 years is significant enough to warrant augmentation of the program in the future."

"The National Tsunami Hazard Mitigation Program has made great progress considering the low level of funding that it has received. The persons participating in the program deserve great praise for their accomplishments under this low level of funding, and funding levels should be increased to meet the demands of program needs."

Because of the consistent reviewer comment of the need for more resources, the NTHMP developed a program at the $4.0M level (Bernard, 2001). This enhanced program was funded beginning in 2002.

Among their suggestions for improvement included:

A. Tsunami inundation maps: Although each state needed to produce maps consistent with state policies, a set of common procedures should be developed to ensure consistency of results from state to state. In response to this suggestion, a set of procedures has been established (González *et al.*, this issue). Also, the budget for mapping was increased by a factor of three to meet the following goals for the next 5 years.

- Complete tsunami inundation maps for 75% of U.S. coastal communities at risk in each state (Alaska, California, Hawaii, Oregon, and Washington).
- Produce evacuation maps that are consistent from state to state for mapped communities.

B. Warning center operations: One reviewer observed that the 2-minutes response by the warning center to the Seattle earthquake in 2001 was due to the time of the earthquake – on Wednesday, 28 February 2001, at 10 a.m. He then inquired what would have been the response at 2 a.m.? This underscored the fact that the NOAA tsunami warning centers have staff in the center during normal working hours and on standby at other times. The reviewer recommended that NOAA staff the warning centers for continuous operations. To address the recommendation, the following goal was established:

- The USGS, NOAA, and state agencies disseminate their automated, reviewed earthquake and tsunami notifications as rapidly as is scientifically and technologically possible. Automated notification of preliminary hypocenter and magnitude should be provided within 2 minutes after receipt of sufficient seismological information at observing networks and reviewed information should be provided in 5 minutes.

C. Other reviewer comments and State needs for a specific warning product of wave forecasts for specific locations using graphics led to the following goals:

- For at least one community per state, issue site- and event-specific forecasts of maximum tsunami flooding depth and inland penetration with an average rms error less than 50%.
- Develop a suite of graphical products for dissemination from NOAA's tsunami warning system.

D. State requirements emerging from state mitigation plans and the local tsunami warning workshop led to the following goals:

- Install evacuation notification system (for example: EMWIN, NOAA Weather Radio, telephone alert, etc.) in 50% of coastal communities in each state.
- Reduce tsunami warning system false alarms by 20%.

E. Mitigation: One reviewer observed that there was lack of clarity about mitigation. The reviewer recommended the use of social science to target preparedness and behavior modification to really mitigate the hazard. As a result of this recommendation, the following goal was established:

- Use social science tools to measure tsunami resilience of communities and the effectiveness of the NTHMP.

F. Five additional mitigation goals were developed by the five States including:

- Designate that 25% of communities at risk in each state are TsunamiReady.
- Ensure that public information is available at all beach access points; ensure that evacuation procedures and maps are in all coastal jurisdiction telephone books/utility bills/school sites/hotels. Display education posters in 75% of coastal water oriented/recreation businesses.
- Develop approved engineering guidance in the FEMA Coastal Construction Manual or other appropriate document that

addresses both high seismic and tsunami loading for use in new construction and retrofitting of existing structures.
- Convince 25% of the potentially threatened businesses to include tsunami components in their business continuity plans.
- Ensure the National Response Plan (NRP) comprehensively addresses tsunami response and recovery.

5. Conclusions

Historically, tsunami mitigation efforts in the United States have followed disaster. The 1946 tragedy in Hawaii resulted in the establishment of the first tsunami warning center. When disaster struck Hawaii again following the 1960 Chilean earthquake, permanent land use changes were incorporated into the Hilo City plan. The 1964 Alaskan tsunami resulted in recognition of the near-source tsunami hazard and the establishment of the ATWC. The NTHMP is unique in having its origins in the recognition of a future hazard and not as a response to recent disaster.

The NTHMP was developed at a time of tight federal budgets and little public or legislative recognition of tsunami hazards. However, through integrated activities of the three efforts of the program – hazard identification, warning guidance, and mitigation products – the tsunami threat to U.S. coastlines has been reduced. The success of the program is the result of personal commitment by Steering Group members. During the first 5 years of the NTHMP funding came through the Congressional add-on process. This required considerable effort to demonstrate the need for tsunami hazard mitigation and its accomplishments on an annual basis to legislators. The Steering Group membership pairing scientists and emergency managers has had a positive impact on the relationship of the tsunami research community to the emergency management community. All states and program elements were required to give progress reports at twice-yearly steering group meetings, stimulating discussion and fostering friendly competition to produce results. The close contacts forged among the Steering Group members and the tsunami community has allowed for a coordinated response to the media when tsunami events occur globally. This was particularly significant in 1998 when a tsunami in Papua New Guinea killed over 2,200 people. Steering group members were able to parlay media interest in the event into raised awareness of tsunami hazards in the U.S.

The program has achieved its initial goals of developing methods for producing scientifically based hazard maps, of establishing a tsunami forecast capability by upgrading the seismic network and establishing a deep ocean tsunameter network, and raising tsunami awareness among the coastal population that will eventually lead to tsunami-resilient communities. These accomplishments have been documented in 99 publications. The NTHMP has been reviewed by a team of outside experts and has developed a set of 14 goals for the next 5 years. The 17 November 2003 tsunami warning was canceled early, avoiding a $68M evacuation in Hawaii. Early indications of the program reflect high benefits for relatively low costs. The goals collectively begin the process of building tsunami-resilient communities that can coexist with tsunami hazards.

Acknowledgements

The author had the privilege of serving as the first chair of the Steering Group from 1996 to 2004. Although engaging discussions were a hallmark of the Group's meetings, the group derived energy from this intrinsic passion. This energy enabled us to "move mountains." The author gratefully acknowledges the dedicated efforts of the present members of the Steering Group, including (in alphabetical order) Jonathan C. Allan, Glenn Bauer, Landry Bernard, Rodney Combellick, George Crawford, Mark Darienzo, Richard Eisner, Frank González, Roger Hansen, Don F. Hoirup, Michael Hornick, Chris Jonientz-Trisler, Laura Kong, Jeff LaDouce, David Oppenheimer, James Partain, Juan M. Pestana, Ervin Petty, George R. Priest, Michael S. Reichle, R. Scott Simmons, Timothy Walsh, Craig Weaver, Brian Yanagi, and Sterling Yong. Also, the author acknowledges the contributions of former Steering Group members, including (in alphabetical order) Lori Dengler, Augustine Furumoto, Richard Hagemeyer, Richard Hutcheon, Don Hull, Hugh Milburn, and Mike Webb.

References

Bernard, E. N.: 1998, Program aims to reduce impact of tsunamis on Pacific states. *Eos, Trans. AGU* **79**(22), 258, 262–263.

Bernard, E. N.: 2001, The U.S. National Tsunami Hazard Mitigation Program Summary. In: *Proceedings of the International Tsunami Symposium 2001 (ITS 2001)* (on CD-ROM), NTHMP Review Session, R-1, Seattle, WA, 7–10 August 2001, pp. 21–27, http://www.pmel.noaa.gov/its2001/.

Bernard, E. N. and González, F. I.: 1994, Tsunami Inundation Modeling Workshop Report (November 16–18, 1993). Technical Report NOAA Tech. Memo. ERL PMEL-100 (PG94-143377), NOAA/Pacific Marine Environmental Laboratory, Seattle, WA.

Blackford, M. and Kanamori H.: 1995, Tsunami Warning System Workshop Report (September 14–15, 1994). Technical Report NOAA Tech. Memo. ERL PMEL-105 (PB95-187175), NOAA/Pacific Marine Environmental Laboratory, Seattle, WA.

Borrero, J., Yalciner, A.C., Kanoglu, U., Titov, V., McCarthy, D. and Synolakis, C.: 2003, Producing tsunami inundation maps: The California experience. In: A. C. Yalciner *et al.* (eds.), *Submarine Landslides and Tsunamis*, Kluwer, Hingham, MA, pp. 315–326.

Crawford, G. L.: 2005, NOAA Weather Radio (NWR) – A Coastal Solution to Tsunami Alert and Notification. *Nat. Hazards* **35**, 163–171 (this issue).

Darienzo, M., Aya, A., Crawford, G., Gibbs, D., Whitmore, P., Wilde, T. and Yanagi, B.: 2005, Local tsunami warning in the Pacific coastal United States. *Nat. Hazards* **35**, 111–119 (this issue).

Dengler, L.: 1998, Strategic implementation plan for tsunami mitigation projects, approved by the Mitigation Subcommittee of the National Tsunami Hazard Mitigation Program, April 14, 1998. Technical Report NOAA Tech. Memo. ERL PMEL-113 (PB99-115552), NOAA/Pacific Marine Environmental Laboratory, Seattle, WA, http://www.pmel.noaa.gov/pubs/PDF/deng2030/deng2030.pdf.

Dengler, L.: 2005, The role of education in the National Tsunami Hazard Mitigation Program. *Nat. Hazards* **35**, 141–153 (this issue).

Eisner, R. K.: 2005, Planning for tsunami: reducing future losses through mitigation. *Nat. Hazards* **35**, 155–162 (this issue).

González, F. I. and Bernard, E. N.: 1993, The Cape Mendocino Tsunami. *Earthquakes and Volcanoes* **23**(3), 135–138.

González, F. I., Bernard, E. N., Meinig, C., Eble, M. C., Mofjeld, H. O. and Stalin, S.: 2005, The NTHMP tsunameter network, *Nat. Hazards* **35**, 25–39 (this issue).

Good, J. W.: 1995, Tsunami Education Planning Workshop, findings and recommendations. NOAA Tech. Memo. ERL PMEL-106 (PB95-195970), NOAA/Pacific Marine Environmental Laboratory, Seattle, WA.

Johnston, D., Paton, D., Crawford, G., Ronan, K., Houghton, B. and Bürgelt, P.: 2005, Measuring tsunami preparedness in Coastal Washington. United States, *Nat. Hazards* **35**, 173–184 (this issue).

Jonientz-Trisler, C., Simmons, R. S., Yanagi, B., Crawford, G., Darienzo, M., Eisner, R., Petty, E. and Priest, G.: 2005, Planning for tsunami-resilient communities. *Nat. Hazards* **35**, 121–139 (this issue).

McCreery, C. S.: 2005, Impact of the National Tsunami Hazard Mitigation Program on operations of the Richard H. Hagemeyer Pacific Tsunami Warning Center, *Nat. Hazards* **35**, 73–88 (this issue).

National Research Council: 2003, *Preventing Earthquake Disasters: The Grand Challenge in Earthquake Engineering: A Research Agenda for the Network for Earthquake Engineering Simulation (NEES)*. National Academic Press, Washington, D.C., 138 pp.

Oppenheimer, D., Bittenbinder, A., Bogaert, B., Buland, R., Dietz, L., Hansen, R., Maloine, S., McCreery, C., Sokolowsky, T., Weaver, C. and Whitmore, P.: 2005, The seismic project of the National Tsunami Hazard Mitigation Program. *Nat. Hazards* **35**, 59–72 (this issue).

Titov, V. V., González, F.I., Eble, M.C., Mofjeld, H. O., Newman, J. C. and Venturato, A. J.: 2005, Real-time tsunami forecasting: challenges and solutions, *Nat. Hazards* **35**, 41–58 (this issue).

Tsunami Hazard Mitigation Federal/State Working Group: 1996, Tsunami Hazard Mitigation Implementation Plan—A Report to the Senate Appropriations Committee. 22 pp., Appendices, http://www.pmel.noaa.gov/tsunami-hazard/hazard3.pdf.

Natural Hazards (2005) 35: 25–39

The NTHMP Tsunameter Network

FRANK I. GONZÁLEZ*, EDDIE N. BERNARD,
CHRISTIAN MEINIG, MARIE C. EBLE, HAROLD O. MOFJELD and
SCOTT STALIN
NOAA/Pacific Marine Environmental Laboratory, Seattle, WA 98115, USA

(Received: 15 December 2003; accepted: 27 April 2004)

Abstract. A tsunameter (soo-NAHM-etter) network has been established in the Pacific by the National Oceanic and Atmospheric Administration. Named by analogy with seismometers, the NOAA tsunameters provide early detection and real-time measurements of deep-ocean tsunamis as they propagate toward coastal communities, enabling the rapid assessment of their destructive potential. Development and maintenance of this network supports a State-driven, high-priority goal of the U.S. National Tsunami Hazard Mitigation Program to improve the speed and reliability of tsunami warnings. The network is now operational, with excellent reliability and data quality, and has proven its worth to warning center decision-makers during potentially tsunamigenic earthquake events; the data have helped avoid issuance of a tsunami warning or have led to cancellation of a tsunami warning, thus averting potentially costly and hazardous evacuations. Optimizing the operational value of the network requires implementation of real-time tsunami forecasting capabilities that integrate tsunameter data with numerical modeling technology. Expansion to a global tsunameter network is needed to accelerate advances in tsunami research and hazard mitigation, and will require a cooperative and coordinated international effort.

Key words: tsunami, tsunameter, tsunami measurement, tsunami warning, tsunami forecast, hazard mitigation, National Tsunami Hazard Mitigation Program

Abbreviations: BPR – bottom pressure recorder, DART – Deep-ocean Assessment and Reporting of Tsunamis, NDBC – National Data Buoy Center, NOAA – National Oceanic and Atmospheric Administration, NTHMP – National Tsunami Hazards Mitigation Program, PMEL – Pacific Marine Environmental Laboratory, PTWC – Pacific Tsunami Warning Center, SIFT – Short-term Inundation Forecast for Tsunamis, SASZ – South American Subduction Zone, TAO – Tropical Atmosphere and Ocean array, USGS – U.S. Geological Survey, WC/ATWC – West Coast and Alaska Tsunami Warning Center

1. Background and Introduction

Just as the worldwide seismometer network has been essential to progress in the field of seismology, a global tsunameter (soo-NAHM-etter) network is critical to the further advancement of tsunami research and hazard mitigation. The U.S. National Tsunami Hazard Mitigation Program (NTHMP),

*Author for correspondence: Tel: +1-206-526-6803; Fax: +1-206-526-6485; E-mail: frank.i.gonzalez@noaa.gov

led by the National Oceanic and Atmospheric Administration (NOAA), has taken a first important step with the development and field-testing of the first generation of reliable tsunameters (Figure 1) and the successful establishment of a Pacific network (Milburn *et al.*, 1996; Meinig *et al.*, 2001; Bernard *et al.*, 2001). The operational network (Figure 2), though currently small, is a powerful catalyst for the revolutionary paradigm shift now underway in tsunami research and forecasting – away from indirect observations and toward direct, high-quality measurements and analyses of the tsunami itself.

Heretofore, tsunami research and operational decisions of NOAA's Pacific Tsunami Warning Center (PTWC) and West Coast and Alaska Warning Center (WC/ATWC) have depended primarily on analyses of seismic information and coastal tide gage measurements. Though valuable, these data are essentially indirect and their interpretation is highly problematic. Seismic data interpretation involves poorly understood seismic/hydrodynamic coupling. Similarly, the interpretation of tide gage data is difficult because of the complex tsunami transformations induced by interaction with shelf, coastline, and harbor features. Furthermore, a tide gage may not survive the impact of the tsunami itself and, if it does survive, the data are not reported until after the tsunami strikes a coastal community. Finally, though coastal tide gages are very useful to warning operations (and extremely valuable in post-event scientific case studies) they

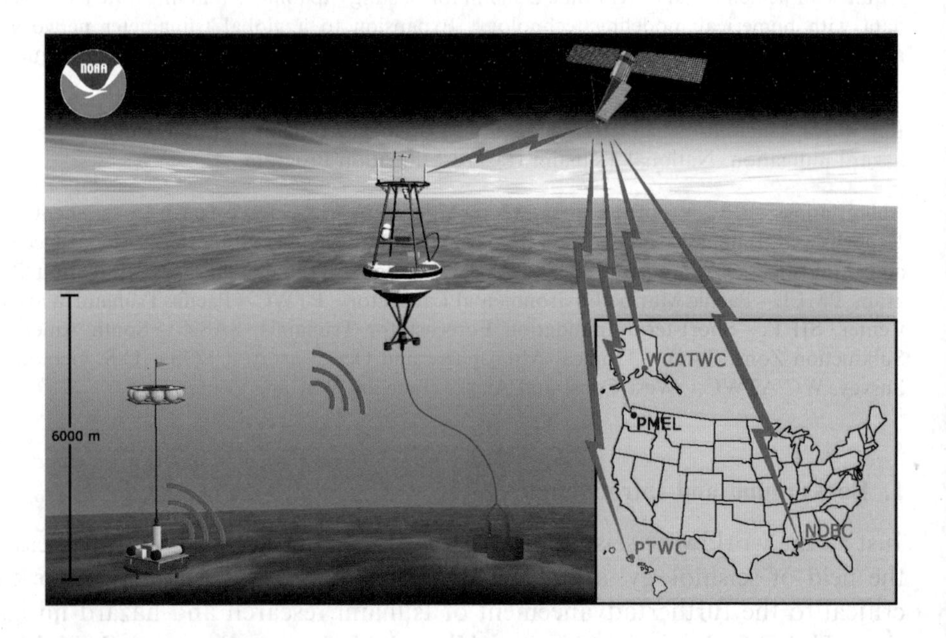

Figure 1. The NOAA tsunameter, illustrating the four major components that had to be integrated into a single system (see text): BPR, acoustic link, surface buoy, and satellite telecommunications.

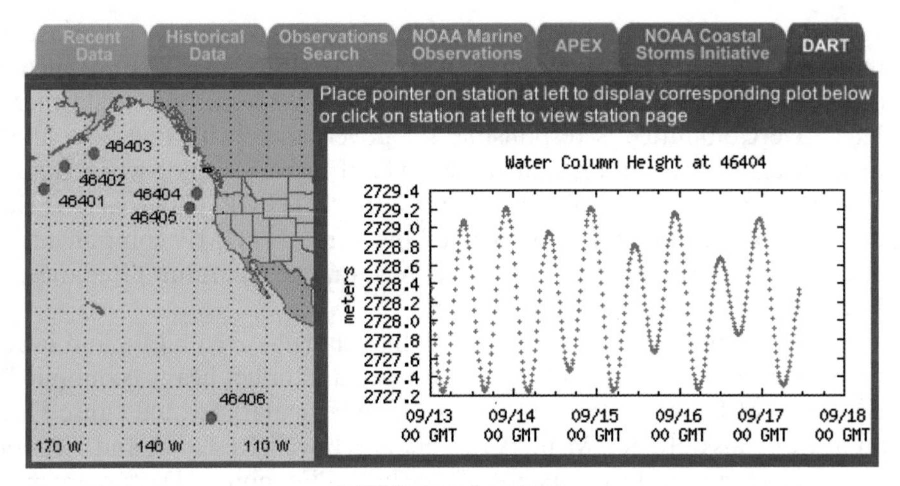

Figure 2. NDBC web page, at URL http://www.ndbc.noaa.gov/dart.shtml. The current NOAA Pacific tsunameter network and the real-time data display are shown. Information on individual stations and relevant reports that can be viewed on-line are also accessible through links at this site.

cannot provide data that are especially important to operational hazard assessment – direct, deep ocean measurements of tsunamis as they propagate from the source to coastal communities.

Engineering advances by NOAA's Pacific Marine Environmental Laboratory (PMEL) have now enabled creation of the NOAA tsunameter, a highly reliable system that acquires and delivers direct tsunami measurements at deep ocean locations between the source and distant communities. This report provides an overview of the research and development effort, the current state of the network, and plans for future technical improvements and expansion.

2. The NTHMP and State-Driven Goals for Warning Improvement

The NTHMP is a partnership of the five Pacific States – Alaska, California, Hawaii, Oregon, and Washington – with NOAA, the U.S. Geological Survey (USGS), the Federal Emergency Management Agency, and the National Science Foundation. NOAA bears primary national responsibility for tsunami warnings and hazard mitigation. Accordingly, the overarching goal of the NTHMP is to reduce the tsunami hazard to U.S. coastal communities. Each State is represented on the 23-member NTHMP Steering Group by at least two individuals, one from the State emergency management agency and another from the State geotechnical agency (the State counterpart to the USGS); similarly, at least two representatives of each Federal Agency are Steering Group members (http://www.pmel.noaa.gov/tsunami-hazard/tsuhaz.htm). An eight-member Executive Committee is responsible for governance, with one vote allotted to each of the five States and three Federal Agencies and, when required, a ninth tie-breaking vote allocated to the NTHMP Chairperson. State-driven goals set priorities in each of the three NTHMP components discussed in this special issue – Hazard Assessment, Warning Guidance, and Mitigation.

False alarms are a serious matter – they damage credibility, and evacuations place citizens at risk of injury or death and inflict heavy economic loss. The State of Hawaii estimated that a single false alarm would cause Hawaii an average loss of $58.2M in 1996 dollars (Hawaii Research and Economic Analysis Division, Department of Business, Economic Development and Tourism, 1996), or about $68M in 2003 dollars. Not surprisingly, therefore, the primary State concern regarding Warning Guidance is improvement of the tsunami warning system and, in particular, the need to "... quickly confirm potentially destructive tsunamis and reduce false alarms." To address this goal, a recommendation was made that the NTHMP "Deploy Tsunami Detection Buoys" that would provide real-time, deep-ocean measurements, thereby improving operational assessments of potentially destructive tsunami impacts and reducing false alarms (Tsunami Hazard Mitigation Federal/State Working Group, 1996).

3. The Engineering Challenge

Development of an operational tsunameter was an extraordinary accomplishment. The task was to design, develop, test, and deploy real-time reporting, deep-ocean instrumentation capable of surviving a hostile ocean environment while performing with the quality and reliability demanded of an operational tsunami warning system on which so many lives depend. The PMEL tsunameter project was initiated to meet this challenge, with the primary requirements listed in Table I as goals that would guide tsunameter design. No such system had ever been developed until the successful effort of the NOAA/PMEL Engineering Development Division.

3.1. THE STRATEGY

As with most effective research and development strategies, "reinventing the wheel" was avoided by an effort to build upon the experience and success of PMEL and others. A number of approaches were explored, but the final basic design consisted of four components: (1) a bottom pressure recorder

Table I. Tsunameter design goals

Reliability and data return	> 80%
Maximum deployment depth	6000 m
Minimum deployment duration	> 1 year
Survivability	Survive N. Pacific winters
Maintenance interval	> 1 year
Sampling interval, internal record	≤15 s
Sampling interval, event reports	≤60 s
Sampling interval, tidal reports	≤15 min
Measurement sensitivity	≤1 mm in 5000 m ($\sim 2 \times 10^{-7}$)
Tsunami data report trigger	a. Automatically by tsunami detection algorithm b. On-demand, by warning center request
Reporting delay	< 2 min
Maximum status report interval	< 6 hours
Cost	< $250K

(BPR) and (2) an acoustic link to (3) a surface buoy equipped with (4) a satellite telecommunications capability (Figure 1).

Three of these four technologies were already in use at PMEL, but had to be modified and integrated into an operational tsunameter. BPR systems with an excellent track record of meeting the tsunameter requirements of reliability, sensitivity, sampling, deployment depth, and deployment duration had been developed earlier by PMEL (Eble and González, 1991; González *et al.*, 1991). Deep-ocean surface buoy technology at PMEL was also well developed, as witness the success of the Tropical Atmosphere and Ocean (TAO) array, the largest deep-ocean array in existence (Hayes *et al.*, 1991; McPhaden, 1993; McPhaden, 1995; McPhaden *et al.*, 1998); a significant challenge had to be overcome in adapting this technology to the needs of a tsunameter network – i.e., development of a buoy and mooring system that would survive the hostile environment of high latitude conditions. Satellite telecommunications had for years been routinely used by PMEL for near real-time data delivery to ground stations from the TAO array, and this technology was also used successfully to deliver real-time seismic data as part of a prototype local tsunami warning system that is still operational in Valparaiso, Chile (Bernard *et al.*, 1988; Bernard, 1991). The remaining component – an acoustic link to provide robust, reliable transmission of BPR data from the seafloor to the surface – represented new, ground-breaking technology, on which much of the development effort focused.

The development, modification, and integration of all four components into a unified tsunameter system, though ultimately successful, proved to be a major engineering challenge. As might be expected, early efforts had to deal with and systematically eliminate a variety of potential problems leading to data dropouts (González *et al.*, 1998). The overall effort, which began in 1996 (Milburn *et al.*, 1996), was remarkable in scope. In time, the enterprise utilized eight different ships for 18 different cruises totaling about 90 days at sea, and the number of participants grew to include more than 25 PMEL engineers, technicians, and scientists, and individuals from more than 85 partner firms and suppliers (Bernard *et al.*, 2001). In September 1997, the first successful deployment of an integrated tsunameter system provided a 3-month record off the Oregon coast, and by 1999 a three-station array was transmitting data from seafloor to desktop with a return rate of 97%, significantly higher than the original goal of 80% presented in Table I (Meinig *et al.*, 2001).

3.2. OPERATIONAL NETWORK PERFORMANCE

A two-year transition period has culminated in the transfer of full operational responsibility to NOAA's National Data Buoy Center (NDBC), including network maintenance and data delivery. PMEL will continue to conduct an active R&D program for future upgrades and enhancements to the tsunameter network. The operational performance of individual tsunameter stations and the network as a whole are presented in Table II in the form of percentage data return rates for the first 8 months of 2003. An excellent NDBC web site provides public access to the real-time data and links to relevant reports that are viewable online: http://www.ndbc.noaa.gov/dart.shtml.

3.3 IDENTIFIED NEEDS – PROGRESS AND PLANS

The formal review of the NTHMP produced comments by the reviewers that identified general recommended improvements to the tsunameter network: an

Table II. Operational percentage data rates of individual tsunameter stations and the entire network for the first 8 months of 2003

	Station						
Month	46401	46402	46403	46404	46405	46406	Network
Jan	99.7	99.7	99.9	22.8	99.9	95.2	86.2
Feb	98.8	99.3	99.4	21.4	100.0	99.6	86.4
Mar	99.7	99.9	99.6	3.0	98.9	94.1	82.5
Apr	99.7	99.3	99.7	13.3	99.9	91.4	83.9
May	99.5	99.3	97.4	6.6	99.7	98.7	83.5
Jun	100.0	99.9	99.2	80.4	99.6	98.2	96.2
Jul	100.0	100.0	99.7	99.9	100.0	99.6	99.9
Aug	100.0	100.0	100.0	87.2	100.0	99.6	97.8
Averages	99.7	99.7	99.4	42.0	99.7	97.0	89.6

increase in the number and geographical coverage of tsunameter network stations; continued improvements in the instrumental technology; forecast methodologies and tools to interpret the data for more effective tsunami warnings. In response to these general recommendations, specific efforts are underway to:

(a) Expand the network and increase the geographical coverage. Though fully operational, the current network is too small. Careful and thorough-going siting studies are needed, but additional stations are clearly required for adequate coverage of all potential tsunami source zones in the Pacific, including the Alaska-Aleutian, Kamchatka, Japan-Kurile, South American, Central American, and Cascadia Subduction Zones. Tsunamis can be highly directional, with a relatively narrow beam of focused energy that could propagate undetected through the network if tsunameters are too widely spaced. Tsunameter spacing of about 200–400 km is required to reliably assess the main energy beam of a tsunami generated by an M8 earthquake (Bernard *et al.*, 2001); beam width decreases with earthquake and tsunami magnitude, with the consequent requirement that tsunameter spacing also be decreased. The length of known tsunamigenic zones in the Pacific is approximately 9000 km, so that the network needs to be expanded to at least 25–50 tsunameter stations.

(b) Develop "on-demand" event mode data delivery. Automatic hourly reports provide "tide mode" data with a 15-minute sampling interval that is capable of resolving low frequency signals with periods of a few hours or more, but not tsunamis. "Event mode" provides data with a 15- to 60-second sampling interval, capable of resolving waveforms in the tsunami period band. Currently, event mode data cannot be acquired by the Tsunami Warning Centers unless an on-board tsunami detection algorithm triggers data transmission. This occurs when a measured wave in the tsunami frequency band exceeds a threshold that is set by software, usually at 3 cm in amplitude (Mofjeld, 1997). The algorithm has performed well, but the disadvantage of this approach is that a station may record an amplitude of less than 3 cm at the low-amplitude fringe of the main energy beam for a tsunami that is, in fact, large and destructive. The warning centers must receive and evaluate all tsunami observations, whether or not their amplitudes exceed 3 cm at a particular station. Such evaluations are essential during an event – it is clearly more desirable to cancel a warning based on real data, rather than on the absence of triggered data. During the early stages of tsunameter development, it was not possible to send a data delivery command through the satellite and acoustic communication links. Recent engineering advances have now made bi-directional satellite communications feasible. In June 2003, a prototype bi-directional tsunameter was deployed 200 nautical miles off the Oregon coast and has been reliably tripped into a high data rate mode from a desktop. Additional engineering development and the establishment of systems to acquire Iridium data at Tsunami Warning Centers will be required before this system becomes operational.

(c) Increase deployment duration. Increasing the servicing interval will lower costs, especially ship time expenses, and reduce the network maintenance effort, thereby facilitating network expansion. The current maintenance cycle is about 1 year for the surface buoy and about 2 years for the ocean bottom unit. A reasonable goal is to lengthen these cycles to 2 and 4 years, respectively.

(d) Develop data interpretation and tsunami forecasting tools. Optimizing the operational value of the tsunameter network to provide accurate, reliable guidance to operational decision-makers requires implementation of a tsunami forecasting system that applies well-established methods for the integration of real-time measurements and numerical modeling. By necessity, NTHMP resources were focused during the first 5 years of the program on the design, development, and testing of the tsunameter hardware and on the establishment of the Pacific network. The network must continue to be improved and expanded, but the NTHMP now requires a parallel effort to exploit and integrate the tsunameter data stream into an accurate, reliable, model-based forecast system to provide real-time predictions of tsunami impacts on threatened communities.

In short, the NTHMP is responding to identified needs by initiating and supporting two major efforts: the design, development, testing, and deployment of an expanded network of the next-generation of tsunameters (Bernard *et al.*, 2001), and the implementation of a tsunami forecasting system to integrate real-time tsunameter data with numerical modeling technology (Titov *et al.*, 2001; Titov and González, 2002; González *et al.*, 2002; González *et al.*, 2005a; Titov *et al.*, this issue).

Next-generation tsunameter design features include extended maintenance intervals of 2 years for the surface buoy and 4 years for the ocean bottom unit, and two-way communication via Iridium satellite telecommunications and acoustic modem for on-demand data delivery. The current network will be expanded to 10 next-generation tsunameter stations by 2008. This includes a tsunameter purchased by Chile, which was deployed on 23 November 2003 near 20°S, 75°W off the Chilean coast at a site approximately 4950 m deep.

Other countries have expressed interest, but none have yet identified the funds needed to establish additional tsunameter stations. By far, the largest expense in establishing a new station and performing the necessary maintenance is ship time, which currently costs about $22K per day. In contrast, the hardware investment is relatively small – about $250K for a new system and $30K per year for maintenance – especially when compared to cable-based systems.

Tsunami forecasting tools will include several redundant methodologies for formal inversion of tsunameter data to produce model-based, site- and event-specific predictions of coastal and inland wave height, inundation depth, and currents. Improvements will be made to the first, basic capability for coastal forecasts developed at WC/ATWC (Whitmore, 2003). More sophisticated methods that were developed at NOAA's Center for Tsunami Inundation Mapping Efforts (Titov *et al.*, this issue) and the University of Hawaii (Wei *et al.*, 2003) will also be implemented. These forecast estimates will be produced and displayed in tabular and graphical form through a graphical user interface as part of the Short-term Inundation Forecast for Tsunamis (SIFT) system (González *et al.*, 2002). Current plans call for an improved coastal forecast capability to be implemented in 2004, followed by implementation of event- and site-specific inland forecast tools over the next few years.

4. The Proven Value

The NOAA tsunameter was developed in response to the high priority assigned by the Pacific States to "... quickly confirm potentially destructive tsunamis and reduce false alarms" (Tsunami Hazard Mitigation Federal/State Working Group, 1996). To this end, even without sophisticated forecasting tools, the immediate value of the network is clear – tsunameter records, especially those acquired directly seaward of the source, can help verify the existence or absence of destructive tsunami energy propagating toward distant communities. Since the network was established, its value has been demonstrated by a number of earthquake events with tsunamigenic potential.

In particular, Table III summarizes six incidents in which tsunameter data have been of assistance in avoiding potential false alarms, including the most recent tsunamigenic earthquake occurrence, on 17 November 2003 at 06:43 UTC. In this case, a warning was issued for Alaska at 07:07, then cancelled at 08:12, shortly after a tsunameter registered a maximum deep-ocean tsunami amplitude of 2 cm (Titov *et al.*, this issue). Costly and potentially hazardous evacuations of Alaskan and Hawaiian coastal communities were thereby averted.

A brief description of the important role of tsunameter data during the event on 11 July 2000, and an overall perspective and judgment on the value of the network has been provided by the Director of the Pacific Tsunami Warning Center (McCreery, 2001):

> "One of these gauges, off Kodiak Island, has already demonstrated its utility by triggering emergency transmissions following a magnitude 6.8 earthquake near Kodiak Island on 11 July 2000. PTWC was able to use these data to quickly confirm that no teletsunami had been generated

Table III. Tsumameter network contribution to operational decisions by NOAA's Tsunami Warning Centers during potentially tsunamigenic events. "Seismic wave induced" signals occur when the ocean bottom pressure sensor is vertically accelerated by passage of the seismic wave

Date – magnitude, time (UTC), location	Tsunameter records	Contribution to operational decisions
11 Jul 2000 – 6.5 M, 01:33 ~70 km SW of Kodiak, AK	No tsunameters were triggered.	Corroborative information for decision not to issue warning. Hawaii Dept. Emerg. Mgt. also requested and received information on tsumameter records (Yanagi, 2000).
10 Jan 2001 – 6.9 M, 16:03 ~110 km SW of Kodiak, AK	Seismic wave induced 3.2 cm signal that triggered tsunameter D157 at 16:11. Subsequent record was tsunami-free.	Tsunameter data allowed PTWC personnel to "... quickly confirm that potentially destructive tsunami waves were not propagating towards Hawaii or the rest of the Pacific" (Goldman, 2001).
5 May 2002 – 6.5 M, 05:37 ~160 km SW of Sand Point, AK	Seismic waves induced signals that triggered three tsunameter stations. Subsequent records were tsunami-free.	Corroborative information for decision not to issue warning.
3 Nov 2002 – 7.9 M, 22:13 ~145 km S of Fairbanks, AK	Seismic waves induced signals that triggered all six tsunameter stations. Subsequent records were tsunami-free.	Corroborative information for decision not to issue warning.
23 Jun 2003 – 7.1 M, 12:13 Near Rat Is., Aleutian Islands	No tsunameters were triggered.	The combination of no trigger at tsunameter D165 with a tsunami-free signal at the Adak coastal gauge, and exercise of the WC/ATWC forecast tool led to early cancellations of the WC/ATWC warning/watch and PTWC Hawaii advisory (McCreery, 2003).

Table III. Continued

Date – magnitude, time (UTC), location	Tsunameter records	Contribution to operational decisions
17 Nov 2003 – 7.5 M, 06:43 ~90 km SW of Amchitka, AK	Seismic waves induced signals that triggered three tsunameter stations. Subsequent records registered maximum deep ocean tsunami amplitudes of 2 cm, 0.5 cm, and < 0.2cm.	07:07 – Alaska warning issued. 07:33–08:03 – Tide gage at Shemya, AK, registers 25 cm maximum. 07:50–08:05 – Tsunameter registers 2 cm maximum. 08:12 – Warning cancelled.

and thus there was no threat to Hawaii. Two more DART[1] gauges sited off the coast of Washington and Oregon would provide Hawaii with timely information about a Cascadia subduction zone event and also measure tsunami waves propagating toward Washington and Oregon from Alaska or even Japan. The sixth gauge, not yet deployed, will go along the equator in the eastern Pacific to provide readings of tsunamis generated in South America as they head toward Hawaii and the West Coast. This gauge would have been useful for more quickly evaluating long range destructive potential of the 23 June 2001 tsunami from Peru. The ultimate utility of the DART gauges won't be realized, however, until their data is incorporated into a tsunami forecasting scheme based on data from numerical tsunami simulations. It is expected that the use of this data, described in more detail below, will lead to a reduction in unnecessary warnings and evacuations and provide better forecasts for levels of tsunami severity."

As this statement notes, the operational value of the tsunameter network will continue to increase as network coverage expands, as warning centers continue to integrate tsunameter network data into their real-time data stream, as SIFT forecast guidance tools continue to be implemented, and as warning center personnel continue to familiarize themselves with and gain confidence in both the tsunameter data and the forecast guidance system.

5. Summary and Conclusions

The NOAA-led U.S. National Tsunami Hazard Mitigation Program has established a tsunameter network in the Pacific operated by NDBC,

[1]Project DART developed the tsunameter.

consisting of six deep-ocean stations located seaward of known tsunamigenic zones. This major engineering accomplishment responds to a State-driven priority for the Warning Guidance component of the NTHMP – i.e., increase the accuracy and reliability of tsunami warnings, to "... quickly confirm potentially destructive tsunamis and reduce false alarms." The network is reliable and the real-time data stream has proven its value to warning center decision-makers during a number of potentially tsunamigenic events. Network improvements are underway – network stations will be increased from the current six to ten by 2008, implementation of real-time tsunami forecasting tools is proceeding, and a next-generation tsunameter is under development that features on-demand data delivery and increased deployment duration and maintenance cycles.

International participation is now needed to speed up expansion and create a global tsunameter network. This network will transform and accelerate advances in tsunami research and hazard mitigation, much as the global seismometer network has had a dramatic impact on the field of seismology. Research aided by tsunameter data includes such basic issues as the degree of nonlinearity, dispersion, and scattering in the deep ocean. Research on the highly nonlinear dynamics of inundation will also benefit. This follows from the specification of more accurate initial conditions for these nonlinear models through both direct tsunameter measurements and improved deep ocean theory; in turn, more accurate initial conditions will help isolate and study the physics of the inundation process to explain discrepancies between observed and simulated runup. Finally, research on tsunami generation will be greatly aided by the acquisition of more near-source tsunameter records. All such research, of course, will improve hazard mitigation programs and products developed by two important components of the NTHMP – Hazard Assessment (González *et al.*, 2005b), and Warning Guidance (Titov *et al.*, this issue).

Japan has deployed real-time reporting BPRs off its coast, using underwater cable technology for power and data transmission (Hirata *et al.*, 2002). This cable approach is effective, but initial costs are prohibitively high (tens of millions of dollars), maintenance and repair is difficult and expensive, and the systems are not easily re-located if required by a change in priorities and/ or scientific understanding of tsunami risk. Chile, with more than 6,000 km of coastline that abuts the South American Subduction Zone (SASZ), is the first country to purchase a NOAA tsunameter, and has now established the first of several planned offshore stations. If a tsunami is generated near a station, it will be detected before it reaches communities at distant points on the long Chilean coast, providing early information and a few extra minutes of warning time that can be critical to reducing fatalities. Furthermore, the station will continue to monitor offshore tsunami activity for the duration of an event, allowing continual assessment of the hazard to coastal residents and, again, reducing casualties. Indeed, all Pacific Rim nations will benefit from the tsunameter station established and maintained by Chile that will enable the early detection, direct measurement, and assessment of the hazard posed to their coastal communities.

More generally, the investment of any nation in tsunameter stations will benefit both itself and other coastal nations that border a common sea or oceanic basin. Additional benefits related to climate and weather could also be realized, with modification of the tsunameter system to provide a platform for meteorological and oceanographic instrumentation. In summary, the mutually beneficial nature of national efforts and investments in a global tsunameter network provides a solid rationale for an internationally cooperative and coordinated program to make such a network a reality.

Acknowledgements

We express our appreciation to the numerous individuals that participated in and contributed to the tsunameter development enterprise, including PMEL engineers, technicians, and scientists, and more than 85 partner firms and suppliers. Cheryl Demers, NDBC, provided the tsunameter network statistics presented in Table II. Ryan Whitney prepared the final manuscript. This work was jointly supported by NOAA and the NTHMP, and is PMEL contribution number 2673.

References

Bernard, E. N.: 1991, Assessment of Project THRUST: Past, present, future. Special Issue on Tsunami Hazard (E.N. Bernard, ed.), *Nat. Hazards* **4**(2,3), 285–292.

Bernard, E. N., Behn, R. R., Hebenstreit, G. T., González, F. I., Krumpe, P., Lander, J. F., Lorca, E., McManamon, P. M., and Milburn, H. B.: 1988, On mitigating rapid onset natural disasters: Project THRUST (Tsunami Hazards Reduction Utilizing Systems Technology). *Eos, Trans. AGU* **69**(24), 649–661.

Bernard, E. N., González, F. I., Meinig, C., and Milburn, H. B.: 2001, Early detection and real-time reporting of deep-ocean tsunamis. In: *Proceedings of the International Tsunami Symposium 2001 (ITS 2001)* (on CD-ROM), NTHMP Review Session, R-6, Seattle, WA, 7–10 August 2001, pp. 97–108. http://www.pmel.noaa.gov/its2001/.

Eble, M. C. and González, F. I.: 1991, Deep-ocean bottom pressure measurements in the northeast Pacific. *J. Atmos. Ocean. Tech.* **8**(2), 221–233.

Goldman, J.: 2001, NOAA Tsunami Buoy "Feels" Alaska Earthquake. http://www.noaa.news.noaa.gov/stories/s560.htm.

González, F. I., Mader, C. L., Eble, M., and Bernard, E. N.: 1991, The 1987–88 Alaskan Bight Tsunamis: Deep ocean data and model comparisons. *Nat. Hazards* **4**(2,3), 119–139.

González, F. I., Milburn, H. M., Bernard, E. N., and Newman, J.: 1998, Deep-ocean assessment and reporting of tsunamis (DART): Brief overview and status report. In:

Proceedings of the International Workshop on Tsunami Disaster Mitigation. Tokyo, Japan, pp. 118–129.

González, F. I., Titov, V., Mofjeld, H. O., and Newman, J. C.: 2002, Project SIFT (Short-term Inundation Forecasting for Tsunamis). In: *Underwater Ground Failures on Tsunami Generation, Modeling, Risk and Mitigation*. Istanbul, Turkey, pp. 221–226.

González, F. I., Titov, V. V., Mofjeld, H. O., Venturato, A., Simmons, S., Hansen, R., Combellick, R., Eisner, R., Hoirup, D., Yanagi, B., Young, S., Darienzo, M., Priest, G., Crawford, G., and Walsh, T.: 2005b, Progress in NTHMP Hazard Assessment. *Nat. Hazards* **35**, 89–110 (this issue).

González, F., Burwell, D., Cheung, K. F., McCreery, C., Mofjeld, H., Titov, V., and Whitmore, P.: 2005a, Far-field tsunami forecast guidance. NOAA Tech. Memo. OAR PMEL, NOAA/PMEL (in preparation).

Hawaii Research and Economic Analysis Division, Department of Business, Economic Development and Tourism: 1996, Methodology for Estimating the Economic Loss of a Tsunami False Alert. Attachment to May 22, 1996 Memorandum from P.I. Iboshi to R.C. Price on "Tsunami Alert Economic Loss Estimation," 6 pp., with 3 Tables.

Hayes, S. P., Mangum, L. J., Sumi, J. P. A., and Takeuchi, K.: 1991, TOGA-TAO: A moored array for real-time measurements in the tropical Pacific Ocean. *Bull. Am. Meteorol. Soc.* **72**(3), 339–347.

Hirata, K., Aoyagi, M., Mikada, H., Kawaguchi, K., Kaiho, Y., Iwase, R., Morita, S., Fujisawa, I., Sugioka, H., Mitsuzawa, K., Suyehiro, K., Kinoshita, H., and Fujiwara, N.: 2002, Real-time geophysical measurements on the deep seafloor using submarine cable in the southern Kurile subduction zone. *IEEE J. Oceanic Eng.* **27**(2), 170–181.

McCreery, C. S.: 2001, Impact of the National Tsunami Hazard Mitigation Program on operations of the Pacific Tsunami Warning Center. In: *Proceedings of the International Tsunami Symposium 2001 (ITS 2001)* (on CD-ROM), NTHMP Review Session, R-7, Seattle, WA, 7–10 August 2001, pp. 109–117. http://www.pmel.noaa.gov/its2001/.

McCreery, C.: 2003, Personal e-mail communication to E. Bernard on June 23, 2003.

McPhaden, M. J.: 1993, TOGA-TAO and the 1991–92 El Niño/Southern Oscillation event. *Oceanography* **6**(2), 36–44.

McPhaden, M. J.: 1995, The tropical atmosphere ocean array is completed. *Bull. Am. Meteorol. Soc.* **76**, 739–741.

McPhaden, M. J., Busalacchi, A. J., Cheney, R., Donguy, J., Gage, K. S., Halpern, D., Ji, M., Julian, P., Meyers, G., Mitchum, G. T., Niiler, P. P., Picaut, J., Reynolds, R. W., Smith, N., and Takeuchi, K.: 1998, The Tropical Ocean Global Atmosphere observing system: A decade of progress. *J. Geophys. Res.* **103**(C7), 14169–14240.

Meinig, C., Eble, M. C., and Stalin, S. E.: 2001, System development and performance of the Deep-ocean Assessment and Reporting of Tsunamis (DART) system from 1997–2001. In: *Proceedings of the International Tsunami Symposium 2001 (ITS 2001)* (on CD-ROM), NTHMP Review Session, R-24, Seattle, WA, 7–10 August 2001, pp. 235–242. http://www.pmel.noaa.gov/its2001/.

Milburn, H. B., Nakamura, A. I., and González, F. I.: 1996, Real-time tsunami reporting from the deep ocean. In: *Proceedings of the Oceans 96 MTS/IEEE Conference*, Fort Lauderdale, FL, pp. 390–394.

Mofjeld, H. O.: 1997, Tsunami Detection Algorithm. http://www.pmel.noaa.gov/tsunami/tda_documentation.html.

Titov, V. V. and González, F. I.: 2002, Modeling solutions for short-term inundation forecasting for tsunamis. In: *Underwater Ground Failures on Tsunami Generation, Modeling, Risk and Mitigation*, NATO Advanced Workshop. Istanbul, Turkey, pp. 227–230.

Titov, V. V., González, F. I., Mofjeld, H. O., and Newman, J. C.: 2001, Project SIFT (Short-term Inundation Forecasting for Tsunamis). In: *Proceedings of the International Tsunami Symposium 2001 (ITS 2001)* (on CD-ROM), Session 7–2, Seattle, WA, 7–10 August 2001, pp. 715–721. http://www.pmel.noaa.gov/its2001/.

Titov, V. V., González, F. I., Eble, M. C., Mofjeld, H. O., Newman, J. C., and Venturato, A. J.: 2005, Real-time tsunami forecasting: Challenges and solutions. *Nat. Hazards* **35**, 41–58 (this issue).

Tsunami Hazard Mitigation Federal/State Working Group: 1996, Tsunami Hazard Mitigation Implementation Plan – A Report to the Senate Appropriations Committee. PDF download at http://www.pmel.noaa.gov/tsunami-hazard/, 22 pp., Appendices.

Wei, Y., Cheung, K. F., Curtis, G. D., and McCreery, C. S.: 2003, Inverse algorithm for tsunami forecasts. *J. Waterw. Port Coast. Ocean Eng.* **129**(3), 60–69.

Whitmore, P. M.: 2003, Tsunami amplitude prediction during events: A test based on previous tsunamis. In: *Science of Tsunami Hazards*, Vol. 21. pp. 135–143.

Yanagi, B.: 2000, Personal e-mail communication to WC/ATWC on July 12, 2000.

Natural Hazards (2005) 35: 41–58

Real-Time Tsunami Forecasting: Challenges and Solutions

VASILY V. TITOV[1,★], FRANK I. GONZÁLEZ[2], E. N. BERNARD[2], MARIE C. EBLE[2], HAROLD O. MOFJELD[2], JEAN C. NEWMAN[3] and ANGIE J. VENTURATO[3]

[1]*University of Washington, Joint Institute for the Study of the Atmosphere and Ocean (JISA0), Box 354235, Seattle, WA 98195-4235, USA;* [2]*NOAA/Pacific Marine Environmental Laboratory (PMEL), 7600 Sand Point Way NE, Seattle, WA 98115-6349, USA;* [3]*University of Washington, Joint Institute for the Study of the Atmosphere and Ocean (JISA0), Box 354235, Seattle, WA 98195-4235, USA*

(Received: 25 September 2003; accepted: 27 April 2004)

Abstract. A new method for real-time tsunami forecasting will provide NOAA's Tsunami Warning Centers with forecast guidance tools during an actual tsunami event. PMEL has developed the methodology of combining real-time data from tsunameters with numerical model estimates to provide site- and event-specific forecasts for tsunamis in real time. An overview of the technique and testing of this methodology is presented.

Key words: tsunami, real-time forecast, tsunami measurement, tsunami model, data assimilation, data inversion, tsunami warning, tsunameters

Abbreviations: BPR – bottom pressure recorder; DARPA – Defense Advanced Research Projects Agency; MOST – Method of Splitting Tsunamis; PMEL – Pacific Marine Environmental Laboratory; NASA – National Aeronautics and Space Administration; NTHMP – National Tsunami Hazard Mitigation Program; NOAA – National Oceanic and Atmospheric Administration; PDC – Pacific Disaster Center; TWC – Tsunami Warning Center

1. Background

The 21 January 2003 Workshop on Far-field Tsunami Forecast Guidance recommended development and implementation of the next generation tools to provide Far-field Tsunami Forecast Guidance. Following this recommendation, the U.S. National Tsunami Hazard Mitigation Program (NTHMP) has funded the development of the tsunami forecast guidance tools for NOAA's Tsunami Warning Centers (TWCs) and emergency managers (NTHMP Steering Group, 2003). The collaborative efforts will combine several tsunami forecast methodologies (Titov *et al.*, 2001;

★Author for correspondence: Tel.: +1-206-526-4536; Fax: +1-206-526-6485; E-mail: vasily.titov@noaa.gov

Wei *et al.*, 2003; Whitmore, 2003) into practical tsunami forecast tools and implement them at TWCs. NOAA's Pacific Marine Environmental Laboratory (PMEL) started systematic research and development efforts to build practical tsunami forecasting tools in 1997 when the Defense Advanced Research Projects Agency (DARPA) funded the Early Detection and Forecast of Tsunami project to develop tsunami hazard mitigation tools for the Pacific Disaster Center (PDC). This work has continued with follow-up grants from the Department of Defense and the National Aeronautics and Space Administration (NASA) and the NTHMP. The results of this effort (Titov and González, 1997; Titov *et al.*, 1999, 2001) are the foundation for the next generation tsunami forecasting tools for the TWCs. This article provides a summary of this research and documents the accomplishments in developing the tsunami forecast tools to date.

2. The Need for Real-Time Tsunami Forecasts

Emergency managers and other officials are in urgent need of operational tools that will provide accurate tsunami forecast as guidance for rapid, critical decisions in which lives and property are at stake. NOAA's TWCs are tasked with issuing tsunami warnings for the U.S. and other nations around the Pacific. Tsunami warnings allow for immediate actions by local authorities to mitigate potentially deadly wave inundation at coastal communities. The more timely and precise the warnings are, the more effective actions can local emergency managers take and the more lives and property can be saved. At present, TWCs personnel face a difficult challenge: to issue tsunami warning based on incomplete and ambiguous data. The initial warning decisions are based on seismic waves as indirect measurements of tsunami generation. Tsunami confirmation by coastal tide gages may arrive too late for timely evacuation measures. This lack of information can lead to a high false alarm rate and ineffective local response to the tsunami warning. Tsunami forecasting tools based on new tsunami measurement technology and the latest modeling techniques can provide crucial additional information and quantitative measures of tsunami impact potential to guide emergency managers during tsunami events.

3. Challenges of Real-Time Forecasts

Tsunami forecasts should provide site- and event-specific information about tsunamis well before the first wave arrives at a threatened community. The only official information forecasted at present is the tsunami arrival time, which is based on earthquake epicenter location determined from seismic waves. The next generation tsunami forecast will provide estimates of all critical tsunami parameters (amplitudes, inundation distances, current velocities, etc.) based on direct tsunami observations. The technical obstacles of achieving this are many, but three primary requirements are accuracy, speed, and robustness.

3.1. ACCURACY

Errors and uncertainties will always be present in any forecast. A practical forecast, however, minimizes the uncertainties by recognizing and reducing possible errors. In the tsunami forecast, measurement and modeling errors present a formidable challenge; but advancements in the science and engineering of tsunamis have identified and researched most of them.

1. *Measurement Error*. Tsunami measurements are always masked by noise from a number of sources: tides, harbor resonance, instrument response, to name a few. Most of the noise can be eliminated from the record by careful consideration of its sources. However, automating noise elimination during real-time assessment presents a serious challenge.
2. *Model Approximation Error*. The physics of tsunami propagation is better understood than that of generation and inundation. For example, landslide generation physics is currently a very active area of research; and comparative studies have demonstrated significant differences in the ability of inundation models to reproduce idealized test cases and/or field observations.
3. *Model Input Error*. Model accuracy can be degraded by errors in (a) the initial conditions set for the sea surface and water velocity, due to inadequate physics and/or observational information, and (b) the bathymetry/topography computational grid, due to inadequate spatial coverage, resolution, and accuracy, including the difficult issues encountered in merging data from different sources.

3.2. SPEED

We refer here to forecast speed as the time taken to make the first forecast product available to an emergency manager for interpretation and guidance. This process involves at least two important, potentially time-consuming, steps:

1. *Data stream to TWC*. Seismic wave data are generally available first, since their propagation velocities are fast (above 2000 m/s). However, finite time is required to interpret these signals in terms of descriptive parameters for earthquakes, landslides, and other potential source mechanisms. Tsunami waves travel much slower (propagation velocities

are around 200 m/s in the deep ocean). In addition, time of at least a quarter of a wave period (when the leading tsunami wave crests) will be needed to incorporate these data into a forecast. Seismic networks are much more dense than tsunami monitoring networks, but inversion algorithms for both are needed to provide source details.

2. *Model simulation speed.* Currently available computational power can provide real-time forecasts, if the time available for forecasting is sufficiently large and the source can be quickly specified. In fact, if powerful parallel computers and/or pre-computed model results are exploited, model execution time can be reduced almost to zero, at least in principle. In practice, of course, there will always be situations for which the source proximity would make it impossible to provide a warning forecast for the closest coasts. But even a late forecast will still provide valuable assessment guidance to emergency managers responsible for critical decisions regarding response, recovery, and search-and-rescue.

3.3. ROBUSTNESS

With lives and property at stake, reliability standards for a real-time forecasting system are understandably high, and the development of such a system is a difficult challenge. On one hand, an experienced modeler can perform a hindcast study and obtain reasonable, reliable results. Such exercises, however, take months to complete, during which multiple runs can be made with variations in the model input and/or the computational grid that are suggested by improved observations. The results are then examined for errors and reasonableness. It is quite another matter to design and develop a robust system that will provide reliable results in real time, without the oversight of an experienced modeler.

4. Technology for Tsunami Forecasting

Recent advances in tsunami measurement and numerical modeling technology can be integrated to create an effective tsunami forecasting system. Neither technology can do the job alone. Observational networks will never be dense because the ocean is vast. Establishing and maintaining monitoring stations is costly and difficult, especially in deep water. Numerical model accuracy is inherently limited by errors in bathymetry and topography and uncertainties in the generating mechanism. But combined, these techniques can provide reliable tsunami forecasts. Here, we review existing modeling and measurement tools used for PMEL's methodology for real-time tsunami forecasting.

4.1. MEASUREMENT

Several real-time data sources are traditionally used for tsunami warning and forecast. They are (1) seismic data to determine source location and source parameters, (2) coastal tide gage data used for direct tsunami confirmation and for tsunami source inversion studies (mostly research studies not in real-time mode), and (3) real-time deep-ocean data from the NTHMP tsunameter network (Gonzalez *et al.*, this issue; Synolakis *et al.*, 1997; The Economist, 2003). Our strategy for the real-time forecasting is to use the deep-ocean measurement as a primary data source for making the tsunami forecast. There are several key features of the deep-ocean data that make it indispensable for the forecast model input:

1. *Rapid tsunami observation.* Since tsunamis propagate with much greater speed in deeper water, the wave will reach the deep-ocean gage much sooner than an equally distant coastal gage. Therefore, a limited number of strategically placed deep-ocean gages can provide advanced tsunami observation for a large portion of the given coastline. This can be illustrated by a simple consideration of the tsunami travel time difference between the tsunameter and the target coast. Consider Hilo as an example of a coastal community. Figure 1 shows contours (thin lines) of the difference between the tsunami travel time to Hilo and to the D125 tsunameter for every point in the Pacific. For example, a tsunami generated anywhere at zero contour would arrive at Hilo and D125 at the same time (zero difference). The thick line is the 3-hour contour, which outlines sources of tsunamis that would arrive at D125 3 hours earlier than at Hilo, leaving enough time for an evacuation decision. The 3-hour contours are also shown for other existing tsunameters (thick broken lines). The envelope of thick contours (hatched area) outlines sources of tsunamis that would be detected by at least one tsunameter in time to decide on evacuation at Hilo. This diagram demonstrates that even a sparse array of existing tsunameters would, in principle, provide timely tsunami detection from most sources around the Pacific for Hilo (and most Hawaiian communities). In practice, however, more tsunameters will be necessary to provide reliable detection of small deep-ocean tsunami signals. Other coastal communities in the U.S. and around the Pacific are not protected as well as Hawaii by existing tsunameters – a much denser array is required even for basic global tsunami forecast system. In addition, a denser array of tsunameters would also decrease the warning time for most coastal communities.
2. *No harbor response.* Tsunameters are placed in deep water in the open ocean where a tsunami signal is not contaminated by local coastal effects. Coastal tide gages, on the other hand, are usually located inside

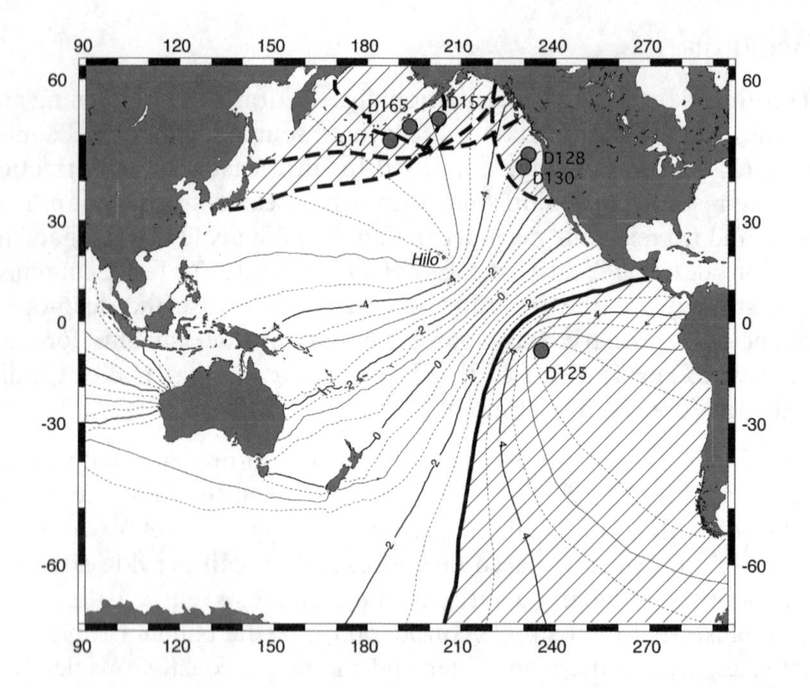

Figure 1. Contours of the time difference between the tsunami arrival at Hilo and at the D125 tsunameter station (thin lines). Thick lines show 3-hour contours of the travel time difference between Hilo and all existing tsunameters (solid line for D125, broken lines for the other tsunameters shown as gray circles). Hatched area outlines sources of tsunamis that reach at least one tsunameter 3 hours before Hilo.

harbors where measurements are subjected to harbor response (Synolakis, 2003). As a result, only part of the tsunami frequency spectrum is accurately measured by coastal gages. In contrast, the tsunameter recording provides "unfiltered" time series with the full spectrum of the tsunami wave.

3. *No instrument response*. The bottom pressure recorder (BPR) of the tsunameter has a very constant frequency response in the tsunami frequency range. Many coastal gages, on the contrary, have complicated and changing frequency characteristics. Since most of the tide gages are designed to measure tides, they often do not perform well in the tsunami frequency band.
4. *Linear process*. The dynamic of tsunami propagation in the deep ocean may be approximated using linear theory because amplitudes are very small compared to the wavelength. This process is relatively well understood, and numerical models of this process are very well developed. The linearity of wave dynamics allows for application of efficient inversion schemes.

4.2. MODELING

The numerical modeling of tsunami dynamics has become a standard research tool in tsunami studies. Modeling methods have matured into a robust technology that has proven to be capable of accurate simulations of historical tsunamis, after careful consideration of field and instrumental data. NOAA's Method of Splitting Tsunami (MOST) numerical model (Titov and Synolakis, 1995, 1997; Titov and González, 1997) is utilized for the development of the tsunami forecasting scheme. This model has been extensively tested against a number of laboratory experiments and was successfully used for simulations of many historical tsunamis (Titov and Synolakis, 1995, 1996, 1997, 1998; Yeh *et al.*, 1995; Bourgeois *et al.*, 1999; Synolakis *et al.*, 2002). Several research groups around the world now use MOST for tsunami mitigation.

The forecast scheme, in contrast to hindcast studies, is a two-step process where numerical models operate in different modes:

1. *Data assimilation mode.* The model is a part of the data assimilation scheme where the model source is adjusted "on-the-fly" by a real-time data stream. The model requirement in this case is similar to hindcast studies: the solution must provide the best fit to the observations. The MOST model has been tested against tsunamis recorded by a deep ocean BPR – the same technology as in the tsunameter system. Figure 2 illustrates one of the early tests. It compares simulated and measured data for the 10 June 1996 Andreanov Is. tsunami. The measurements have very high noise-to-signal ratio (e.g., tsunami amplitude is smaller than low-frequency noise at AK70). Nevertheless, the computed tsunami signal compares well with the recorded leading tsunami wave. Even the tails of tsunami records (which contain reflections from various coastlines) are simulated reasonably well for amplitudes and frequency. The good agreement confirms that the model captures the basic physics of the process and is able to reproduce the data used for the forecast.
2. *Forecast mode.* The model uses the simulation scenario obtained in the first step to extend the simulation to locations where measured data is not available, i.e., providing the forecast. It is difficult to fully assess the forecast potential of a particular model, since the quality and accuracy of the prediction will always depend on the scenario chosen by the data assimilation step. Accurate simulation of the near-shore tsunami dynamics and inundation are especially important. As a partial test of inundation forecast capability of the MOST model, the simulation of the 1993 Hokkaido-Nansei-Oki tsunami has been compared with an independent dataset. The model scenario of this event is based on the field survey data (Takahashi, 1996). An independent, much denser

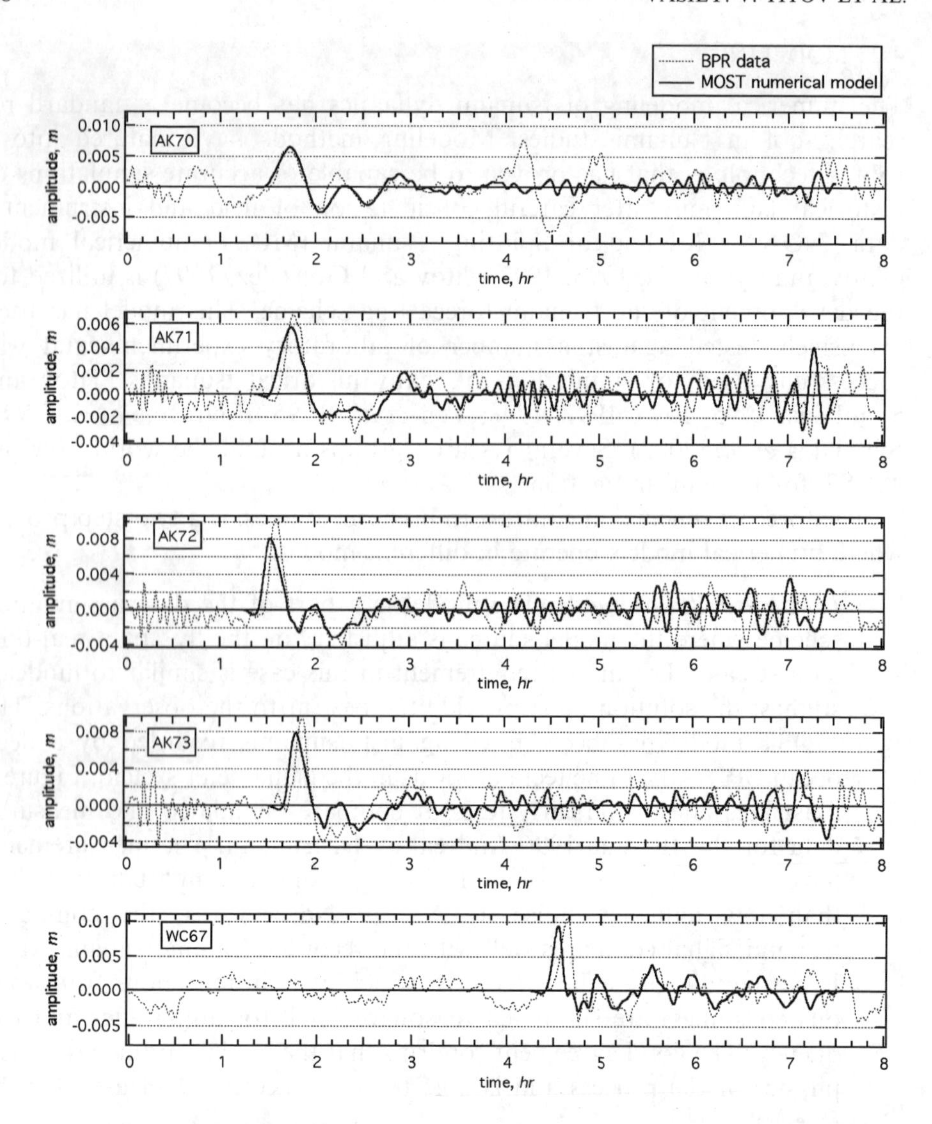

Figure 2. Comparison of the 1996 Andreanov Is. tsunami propagation model (solid line) with the deep-ocean BPR data (dotted line). Locations of the BPRs are shown in Figure 4.

dataset of tsunami inundation distances and heights have been obtained at PMEL from stereo photography data of Okushiri Island. Figure 3 shows a comparison of the original MOST simulation (Titov and Synolakis, 1997) with the new stereo data. Inundation values are compared for the west coast of Okushiri Island, where the highest runup was measured for this event. The MOST runup and inundation estimates compare well with both stereo and field data.

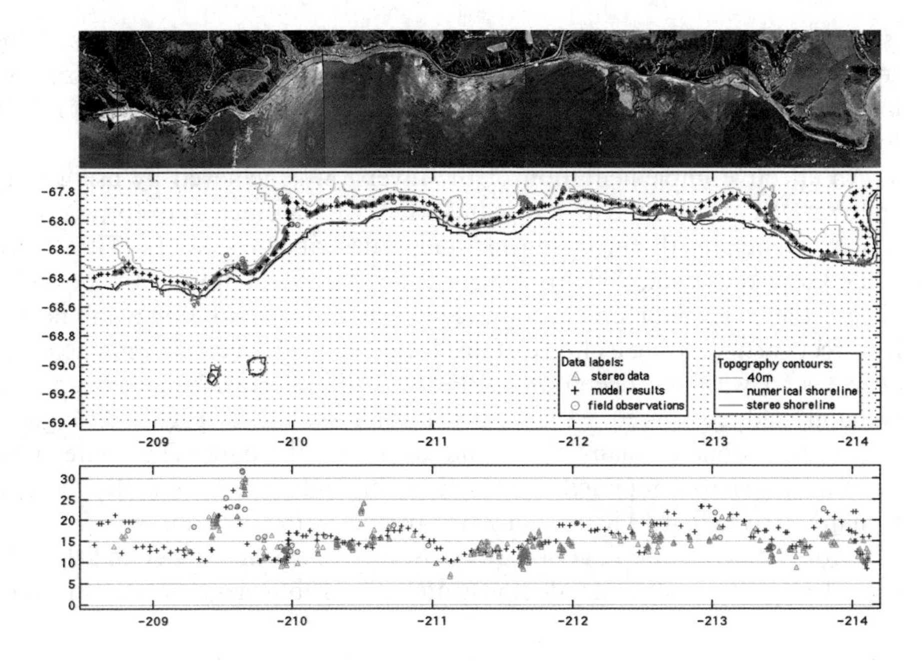

Figure 3. Comparison of the 1993 Okushiri tsunami inundation model (crosses) with field observations (circles) and stereo photo data (triangles). Top frame shows an aerial photo of the modeled area used for the stereo analysis of the inundation data. Middle frame illustrates the numerical grid used for the simulation of the same area (dots are computational nodes, contours show topography data) and compares inundation distances. Bottom frame compares maximum vertical runup for the same shoreline locations.

4.3. DATA ASSIMILATION AND INVERSION

An effective tsunami forecast scheme would automatically interpret incoming real-time data to develop the best model scenario that fits this data. This is a classical inversion problem, where initial conditions are determined from an approximated solution. Such problems can be successfully solved only if proper parameters of the initial conditions are established. These parameters must effectively define the solution, otherwise the inversion problem is ill-posed.

Indeed, several parameters describe a tsunami source commonly used for tsunami propagation simulations (location, magnitude, depth, fault size, and local mechanism). Choosing the subset of those parameters that control the deep-ocean tsunami signal is the key to developing a useful inversion scheme for tsunameter data. A sensitivity study has been conducted to explore this problem. Titov *et al.* (1999, 2001) have studied the sensitivity of far-field data to different parameters of commonly used tsunami sources. The results showed that source magnitude and location essentially define far-field tsunami signals for a wide range of subduction zone earthquakes. Other source parameters have secondary influence and can be ignored during the inversion. This result substantially reduces the size of the inversion problem for the deep-ocean data.

An effective implementation of the inversion is achieved by using a discrete set of Green's functions (ocean surface displacements) to form a model source. Details of the inversion method is described elsewhere (Titov *et al.*, 2003; González *et al.*, 2003b). In short, the algorithm chooses the best fit to given tsunameter data among a limited number of unit solution combinations by direct sorting, using a choice of misfit functions. This inversion scheme has been tested with the deep-ocean BPR records of the 1996 Andreanov Is. tsunami and compared with earlier results shown in Figure 2. Figure 4 demonstrates one of many tests conducted with the data. The figure shows the model scenario obtained by inverting only data from one BPR where the tsunami arrives first (AK72). Only one period of the tsunami wave record is used for the inversion. A good comparison between the model and the BPR data from other stations demonstrates the robustness of the inversion scheme.

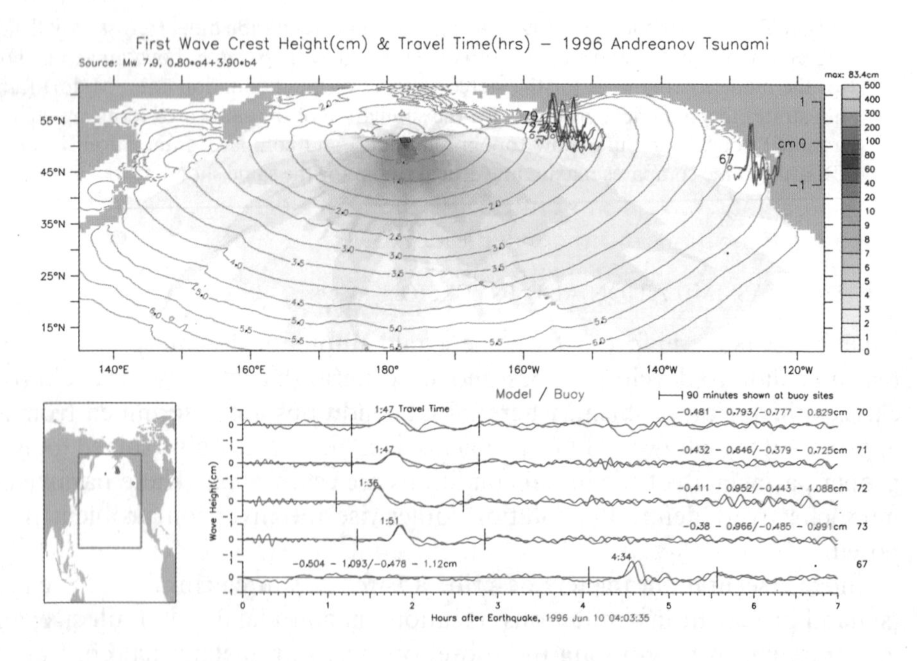

Figure 4. Screenshots of the offshore forecast tool. Results of BPR data inversion for 1996 Andreanov Is. tsunami. Top frame shows the source inferred by the inversion (black rectangles show unit sources' fault plains), maximum computed amplitudes of tsunami from this source (filled colored contours), travel time contours in hours after earthquake (solid lines), and locations of the BPRs. Bottom frame shows a reference map (left) and comparison of the model (blue) and BPR data (magenta).

5. PMEL Methodology for Tsunami Forecasting

The previous discussion suggests that the critical components of tsunami forecasting technology exist now that could provide rapid, usably accurate forecasts of the first few waves. Various ideas for real-time tsunami forecast methods have been discussed in the literature, most suggesting usage of seismic data (e.g., Izutani and Hirasawa, 1987; Shuto *et al.*, 1990). Japan has developed and implemented a local tsunami amplitude forecast system based on the seismic data and interpolation of pre-computed coastal amplitudes (Tatehata, 1997). Without data assimilation from direct tsunami observations, however, such schemes are susceptible to large errors of seismic source estimates. Methods that discuss use of tsunami amplitude data are often difficult to implement for arbitrary tsunamis Pacific-wide (e.g., Koike *et al.*, 2003). PMEL has developed a practical forecast system that combines real-time seismic and tsunami data with a forecast database of pre-computed scenarios. Later waves could also be usefully forecasted by processing real-time tsunami data with a statistical/empirical model (Mofjeld *et al.*, 2000). Implementation of this technology requires integration of these components into a unified, robust system.

5.1. LINEAR PROPAGATION MODEL DATABASE FOR UNIT SOURCES

The source sensitivity study (Titov *et al.*, 1999) has established that only a few source parameters are critical for the far-field tsunami characteristics, namely the location and the magnitude (assuming some typical mechanism for the displacement). Therefore, a discrete set of unit sources (Green's functions) can provide the basis for constructing a tsunami scenario that would simulate a given tsunameter data. Numerical solutions of tsunami propagation from these unit sources, when linearly combined, provide arbitrary tsunami simulation for the data assimilation step of the forecast scheme.

This principle is used to construct a tsunami forecast database of pre-computed propagation solutions for unit sources around the Pacific (Figure 5). Titov *et al.* (1999) described the process of defining the unit sources. Presently, the database contains 246 model scenarios for unit sources that cover historically most active subduction zones around the Pacific. The database stores all simulation data for each unit solution, including amplitudes and velocities for each offshore location around the Pacific. Thus, data assimilation can be completed without additional time-consuming model runs. The methodology also provides the offshore forecast of tsunami amplitudes and all other wave parameters around the Pacific immediately after the data assimilation is complete.

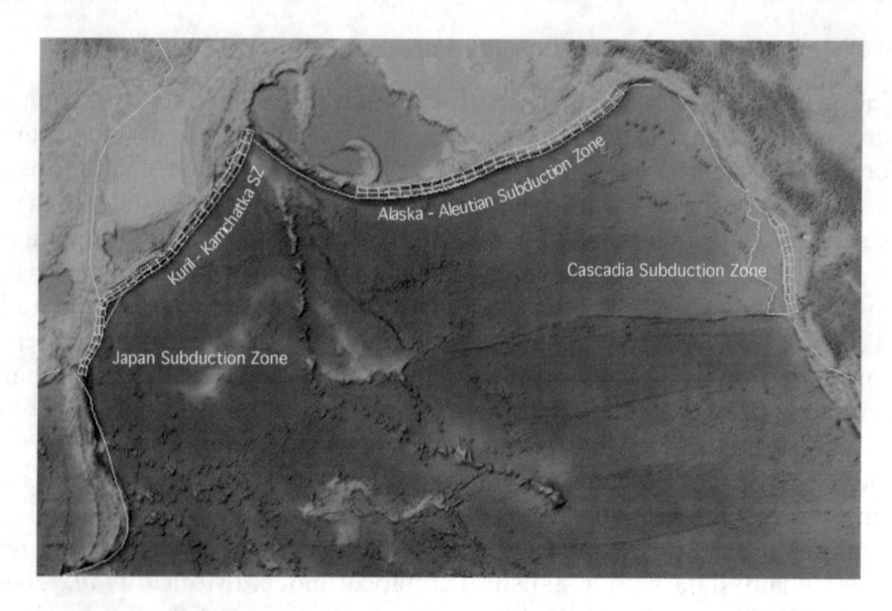

Figure 5. North Pacific details of the Pacific-wide forecast model database. Bathymetric data for the database computation is shown as a shaded relief map. White rectangles show fault planes for the unit sources included in the database. Major plate boundaries are shown as white lines.

5.2. SOURCE CORRECTION USING TSUNAMETER

The previously described inversion algorithm is implemented to work with the forecast database. It combines real-time tsunameter data of offshore amplitude with the simulation database to improve accuracy of an initial offshore tsunami scenario.

5.3. INUNDATION ESTIMATES WITH NON-LINEAR MODEL

Once the offshore scenario is obtained, the results of the propagation model are used for the site-specific inundation forecast. Tsunami inundation is a highly nonlinear process. Therefore, linear combinations of different inundation runs cannot be combined to obtain a valid solution. A high-resolution $2 + 1$ inundation model (Titov and Synolakis, 1998) is run to obtain a local inundation forecast. Data input for the inundation computations are the results of the offshore forecast – tsunami parameters (wave heights and depth-averaged velocity) along the perimeter of the inundation computation area. The forecast inundation model can be optimized to obtain local forecasts within minutes on modern computers.

Nevertheless, obtaining inundation estimates for many communities simultaneously can take too much time. We are considering different approaches to reduce the inundation forecast time, including using parallel supercomputers and/or distributed computation of local inundation via a web interface. Simplified methods of inundation estimation are also being considered for fast preliminary estimates of coastal amplitudes, such as one-dimensional runup estimates (one spatial dimension), analytical extrapolation of the offshore values to the coast, and others.

In summary, to forecast inundation from early tsunami waves, seismic parameter estimates and tsunami measurements are used to sift through a pre-computed generation/propagation forecast database and select an appropriate (linear) combination of scenarios that most closely matches the observational data. This produces estimates of tsunami characteristics in deep water which can then be used as initial conditions for a site-specific (non-linear) inundation algorithm. A statistical methodology has been developed to forecast the maximum height of later tsunami waves that can threaten rescue and recovery operations. The results are made available through a user-friendly interface to aid hazard assessment and decision making by emergency managers. The MOST model performed computations of generation/propagation scenarios for the forecast database. The non-linear 2 + 1 high-resolution model will provide the inundation forecasts.

6. Testing Tsunami Forecasting Methodology

The limited number of deep-ocean tsunami records do not include tsunamis that have been destructive or caused inundation to the U.S. coasts. However, there are several events that were recorded by both deep-ocean and coastal gages. The forecast methodology has been tested against three such tsunamis. The 10 June 1996 Andreanov Is. (Tanioka and González, 1998) and 4 October 1994 Kuril Is. (Yeh *et al.*, 1995) events were recorded by several research BPRs (without real-time data transmission) at similar locations offshore of Alaska and the U.S. West Coast. The offshore model scenario for the Andreanov Is. event was obtained from the forecast database by inverting data from just one BPR as described earlier (Figure 4). The inversion of the Kuril Is. data was done using all five BPR recordings; the results are shown in Figure 6.

The 17 November 2003 Rat Is. tsunami provided the most comprehensive test for the forecast methodology. The Mw 7.8 earthquake on the shelf near Rat Islands, Alaska generated a tsunami that was detected by three tsunameters located along the Aleutian Trench – the first tsunami detection by the newly developed real-time tsunameter system. These real-time data combined with the model database were then used to produce the real-time model tsunami forecast. For the first time, tsunami model predictions were obtained during the tsunami propagation, before the waves had reached many coastlines. The initial offshore forecast was obtained immediately after

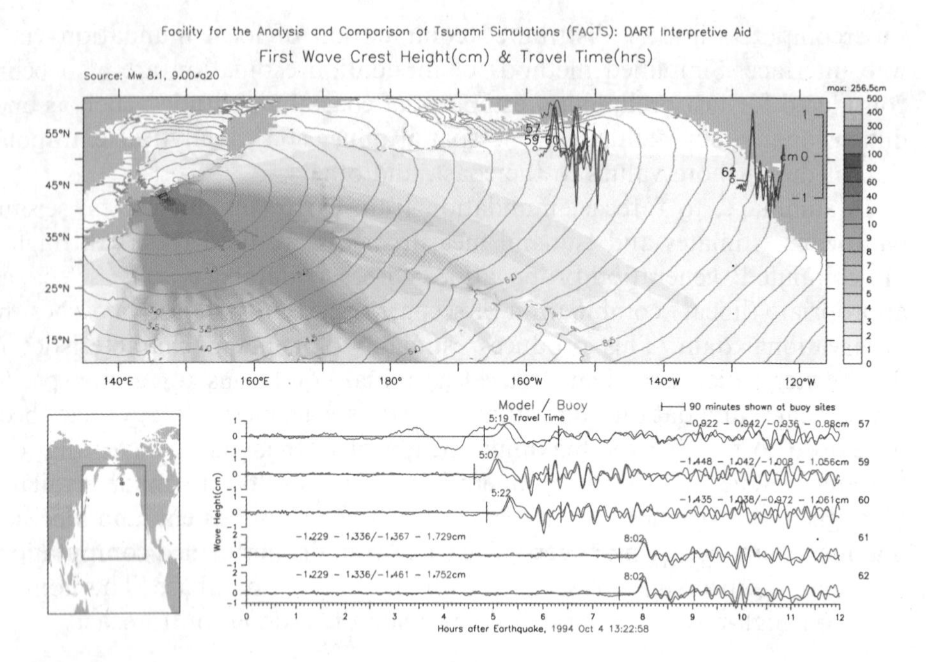

Figure 6. Offshore forecast for the 1994 Kuril Island tsunami. Notations are the same as in Figure 4.

preliminary earthquake parameters (location and magnitude Ms = 7.5) became available from the West Coast/Alaska TWC (about 15–20 minutes after the earthquake). The model estimates provided expected tsunami time series at tsunameter locations. When the closest tsunameter (Sta. 46401-D171) recorded the first tsunami wave, the model predictions were compared with the deep-ocean data and the adjusted forecast was produced immediately, about 1 hour 20 minutes after the earthquake (Figure 7). This adjusted model not only correctly predicted the tsunami records at other locations, it also provided a better estimate of the earthquake magnitude (Mw = 7.7 – 7.8), which was confirmed later by seismic analysis from USGS (Mw = 7.8; National Earthquake Information centre (NEIC), 2003) and the Harvard Seismology Group (Mw = 7.7). The forecast was done in a test mode and was not a part of the TWC operation, but it provided a genuine test of PMEL's forecast method. When implemented, such a forecast will be obtained even faster and would provide enough lead time for potential evacuation or warning cancellation for Hawaii and the U.S. West Coast.

These offshore model scenarios were then used as input for the high-resolution inundation model for Hilo Bay. The model computed tsunami dynamics on several nested grids, with the highest spatial resolution of 30 meters inside Hilo Bay (Figure 8). Neither tsunami produced inundation at Hilo, but all recorded nearly half a meter (peak-to-trough) signal at Hilo

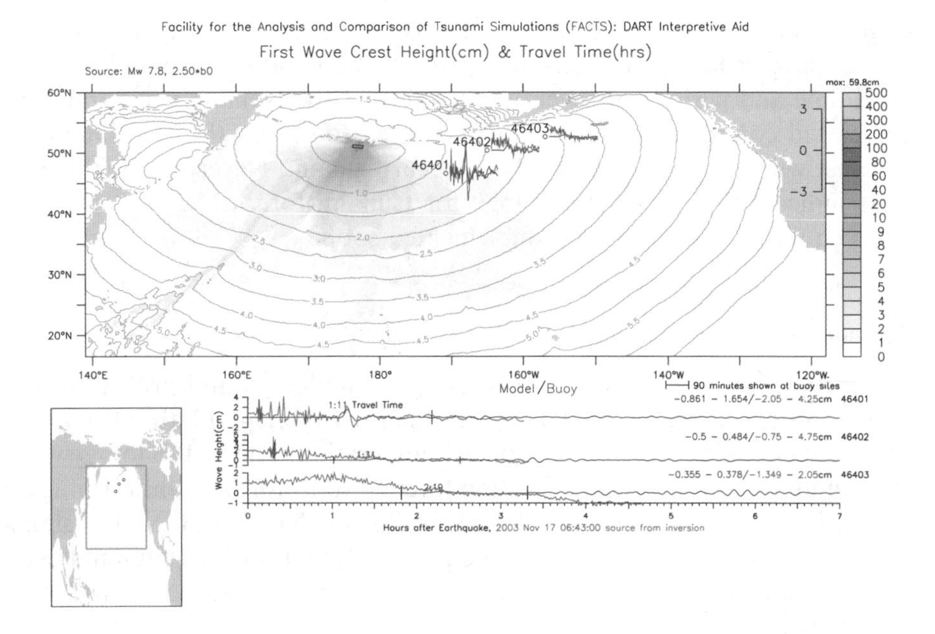

Figure 7. Offshore forecast for the 2003 Rat Island tsunami. Notations are the same as in Figure 4.

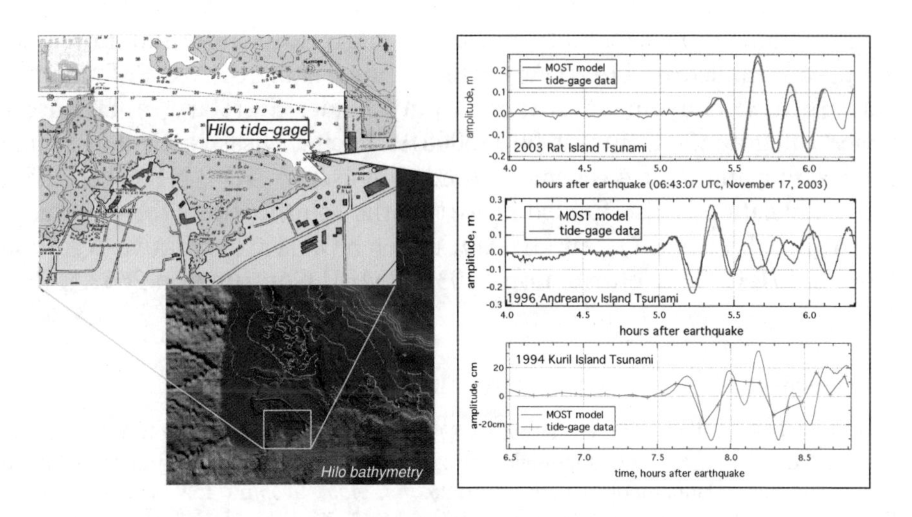

Figure 8. Coastal forecast at Hilo, HI for 2003 Rat Island (top), 1996 Andreanov Is. (middle) and 1994 Kuril Is. (bottom) tsunamis. Left frame shows location of Hilo tide-gage (top map) and digital elevation data for the high-resolution inundation computation (bottom map). Right frame shows comparison of the forecasted (red line) and measured (blue line) gage data.

gage. Model forecast predictions for this tide gage are compared with observed data in Figure 8. The comparison demonstrates that amplitudes, arrival time and periods of several first waves of the tsunami wave train were forecasted correctly. More tests are required to ensure that the inundation forecast will work for every likely-to-occur tsunami. Nevertheless, these first tests indicate that the methodology for tsunami forecast works and useful tools could be developed and implemented soon.

7. Summary

This article describes tsunami forecasting methodology and prototype modeling tools developed by the Pacific Marine Environmental Laboratory of the National Oceanic and Atmospheric Administration. The methodology will be the foundation of the next generation forecast tools for tsunami warning and mitigation that are being developed in close collaboration with Tsunami Warning Centers and academia. The new tools will provide site- and event-specific forecast of tsunami amplitudes for the entire Pacific to assist emergency managers during tsunami warning and mitigation procedures.

Acknowledgements

The development of the tsunami forecasting methodology and supporting research has been funded by the NTHMP and by contracts to develop PDC tsunami mitigation capability, including Early Detection and Forecast of Tsunamis, funded by DARPA; Offshore Forecasting of AASZ Tsunamis in Hawaii, funded by the Assistant Deputy Under Secretary of Defense for Space Integration, Contract NM8600019; Pacific Disaster Center Tsunami Forecasting and Risk Assessment, funded by NASA, RFP NRA-98-OES-13. This publication is also funded in part by the Joint Institute for the Study of the Atmosphere and Ocean (JISAO) under NOAA Cooperative Agreement No. NA17RJ1232, Contribution #1037. PMEL Contribution 2662.

References

Bourgeois, J., Petroff, C., Yeh, H., Titov, V., Synolakis, C., Benson, B., Kuroiwa, J., Lander, J., and Norabuena, E.: 1999, Geologic setting, field survey and modeling of the Chimbote, northern Peru, tsunami of 21 February 1996. *Pure Appl. Geophys.* **154**(3/4), 513–540.

González, F. I., Bernard, E. N., Meinig, C., Eble, M. C., Mofjeld, H. O., and Stalin, S.: 2005, The NTHMP tsunameter network. *Nat. Hazards* **35**, 25–39 (this issue).

González, F. I., Titov, V. V., Avdeev, A. V., Bezhaev, A. Yu., Lavrentiev, M. M., Jr., and Marchuk, An. G.: 2003b, Real-time tsunami forecasting: Challenges and solutions. In: *Proceedings of the International Conference on Mathematical Methods in Geophysics–2003*. Novosibirsk, Russia, pp. 225–228, ICM&MG Publisher.

Izutani, Y. and Hirasawa, T.: 1987, Rapid estimation of fault parameters for near-field tsunami warning. *Nat. Disaster Sci.* **9**, 99–113.

Koike, N., Kawata, Y., and Imamura, F.: 2003, Far-field tsunami potential and a real-time forecast system for the Pacific using the inversion method. *Nat. Hazards* **29**, 423–436.

Mofjeld, H. O., González, F. I., Bernard, E. N., and Newman, J. C.: 2000, Forecasting the heights of later waves in Pacific-wide tsunamis. *Nat. Hazards* **22**, 71–89.

National Earthquake Information Center (NEIC): 2003, *Poster of the Rat Islands, Alaska Earthquake of 17 November 2003–Magnitude 7.8.* URL: http://neic.usgs.gov/neis/poster/2003/20031117.html.

NTHMP Steering Group: 2003, *Summary Report of the Tsunami Hazard Mitigation Steering Group.* FY 03 Budget Meeting, 28–29 May 2003, Seattle, Washington (http://www.pmel.noaa.gov/tsunami-hazard/).

Shuto, N., Goto, G., and Imamura, F.: 1990, Numerical simulation as a means of warning for near-field tsunamis. In: *2nd UJNR Tsunami Workshop*. Honolulu, Hawaii, pp. 133–152.

Synolakis, C. E.: 2003, Tsunami and Seiches. In: W.-F. Chen and C. Scawthorn (eds.), *Earthquake Engineering Handbook*. CRC Press, pp. 9–1—9–90.

Synolakis, C. E., Bardet, J. P., Borrero, J., Davis, H., Okal, E., Sylver, E., Sweet, J., and Tappin, D.: 2002, Slump origin of the 1998 Papua New Guinea tsunami. *Proc. R. Soc. Lond. A* **458**(2020), 763–789.

Synolakis, C. E., Liu, P., Yeh, H., and Carrier, G.: 1997, Tsunamigenic seafloor deformations. *Science* **278**(5338), 598–600.

Takahashi, T.: 1996, Benchmark problem 4. The 1993 Okushiri tsunami–Data, conditions and phenomena. In: Yeh, H., Liu, P., and C. Synolakis (eds.), *Long Wave Runup Models.* Singapore, World Scientific Publishing Co. Pte. Ltd., pp. 384–403.

Tanioka, Y. and González, F. I., 1998, The Aleutian earthquake of June 10, 1996 (Mw 7.9) ruptured parts of both the Andreanov and Delarof segments. *Geophys. Res. Lett.* **25**(12), 2245–2248.

Tatehata, H.: 1997, The new tsunami warning system of the Japan Meteorological Agency. In: G. Hebenstreit (ed.), *Perspectives of Tsunami Hazard Reduction.* Kluwer, pp. 175–188.

The Economist: 2003, *The Next Big Wave.* 14 August 2003.

Titov, V. and González, F. I. : 1997, *Implementation and Testing of the Method of Splitting Tsunami (MOST) model.* Technical Report NOAA Tech. Memo. ERL PMEL-112 (PB98-122773), NOAA/Pacific Marine Environmental Laboratory, Seattle, WA.

Titov, V. V., González, F. I., Lavrentiev, M. M., Bezhaev, A. Yu., Marchuk, An. G., and Avdeev, A. V.: 2003, Assessing tsunami magnitude for inundation forecast. In: *Program and Abstracts of IUGG 2003*, part B. Sapporo, Japan: IUGG 2003 Local Organizing Committee, p. B153.

Titov, V. V., Mofjeld, H. O., González, F. I., and Newman, J. C.: 1999, *Offshore Forecasting of Alaska-Aleutian Subduction Zone Tsunamis in Hawaii.* NOAA Tech. Memo ERL PMEL-114, NOAA/Pacific Marine Environmental Laboratory, Seattle, WA.

Titov, V. V., Mofjeld, H. O., González, F. I., and Newman, J. C.: 2001, Offshore forecasting of Alaskan tsunamis in Hawaii. In: G. T. Hebenstreit (ed.), *Tsunami Research at the End of a Critical Decade*. Birmingham, England, Kluwer Acad. Pub., Netherlands, pp. 75–90.

Titov, V. V. and Synolakis, C. E.: 1995, Modeling of breaking and nonbreaking long wave evolution and runup using VTCS-2. *J. Waterw. Port Coast. Ocean Eng.* **121**(6), 308–316.

Titov, V. V. and Synolakis, C. E.: 1996, Numerical modeling of 3-D long wave runup using VTCS-3. In: Yeh, H., Liu, P., and Synolakis, C. (eds.), *Long Wave Runup Models.* Singapore, World Scientific Publishing Co. Pte. Ltd., pp. 242–248.

Titov, V. V. and Synolakis, C. E.: 1997, Extreme inundation flows during the Hokkaido-Nansei-Oki tsunami. *Geophys. Res. Lett.* **24**(11), 1315–1318.

Titov, V. V. and Synolakis, C. E.: 1998, Numerical modeling of tidal wave runup. *J. Waterw. Port Coast. Ocean Eng.* **124**(4), 157–171.

Wei, Y., Cheung, K. F., Curtis, G. D., and McCreery, C. S.: 2003, Inverse algorithm for tsunami forecasts. *J. Waterw. Port Coast. Ocean Eng.* **129**(3), 60–69.

Whitmore, P. M.: 2003, Tsunami amplitude prediction during events: A test based on previous tsunamis. In: *Science of Tsunami Hazards*, Vol. 21. pp. 135–143.

Yeh, H., Titov, V. V., Gusiakov, V., Pelinovsky, E., Khramushin, V., and Kaistrenko, V.: 1995, The 1994 Shikotan earthquake tsunami. *Pure Appl. Geophys.* **144**(3/4), 569–593.

Natural Hazards (2005) 35: 59–72

The Seismic Project of the National Tsunami Hazard Mitigation Program

DAVID H. OPPENHEIMER[1]*, ALEX N. BITTENBINDER[2], BARBARA M. BOGAERT[2], RAYMOND P. BULAND[2], LYNN D. DIETZ[1], ROGER A. HANSEN[3], STEPHEN D. MALONE[4], CHARLES S. McCREERY[5], THOMAS J. SOKOLOWSKI[6], PAUL M. WHITMORE[6] and CRAIG S. WEAVER[7]

[1]*U.S. Geological Survey, Menlo Park, CA, USA;* [2]*U.S. Geological Survey, Golden, CO, USA;* [3]*University of Alaska, Fairbanks, AK, USA;* [4]*University of Washington, Seattle, WA, USA;* [5]*Richard H. Hagemeyer Pacific Tsunami Warning Center, Ewa Beach, HI, USA;* [6]*West Coast & Alaska Tsunami Warning Center, Palmer, AK, USA;* [7]*U.S. Geological Survey, Seattle, WA, USA*

(Received: 4 September 2003; accepted: 19 April 2004)

Abstract. In 1997, the Federal Emergency Management Agency (FEMA), National Oceanic and Atmospheric Administration (NOAA), U.S. Geological Survey (USGS), and the five western States of Alaska, California, Hawaii, Oregon, and Washington joined in a partnership called the National Tsunami Hazard Mitigation Program (NTHMP) to enhance the quality and quantity of seismic data provided to the NOAA tsunami warning centers in Alaska and Hawaii. The NTHMP funded a seismic project that now provides the warning centers with real-time seismic data over dedicated communication links and the Internet from regional seismic networks monitoring earthquakes in the five western states, the U.S. National Seismic Network in Colorado, and from domestic and global seismic stations operated by other agencies. The goal of the project is to reduce the time needed to issue a tsunami warning by providing the warning centers with high-dynamic range, broadband waveforms in near real time. An additional goal is to reduce the likelihood of issuing false tsunami warnings by rapidly providing to the warning centers parametric information on earthquakes that could indicate their tsunamigenic potential, such as hypocenters, magnitudes, moment tensors, and shake distribution maps. New or upgraded field instrumentation was installed over a 5-year period at 53 seismic stations in the five western states. Data from these instruments has been integrated into the seismic network utilizing Earthworm software. This network has significantly reduced the time needed to respond to teleseismic and regional earthquakes. Notably, the West Coast/Alaska Tsunami Warning Center responded to the 28 February 2001 Mw 6.8 Nisqually earthquake beneath Olympia, Washington within 2 minutes compared to an average response time of over 10 minutes for the previous 18 years.

Key words: tsunami warning, earthquake monitoring, seismic network, seismic instrumentation, earthquake, tsunami

* Author for correspondence: 345 Middlefield Road, MS 977, Menlo Park, CA 94025, USA. Tel: +1-650-329-4792; Fax: +1-650-329-4732; E-mail: oppen@usgs.gov

Abbreviations: AEIC – Alaska Earthquake Information Center, EMWIN – Emergency Managers Weather Information Network, HVO – Hawaii Volcano Observatory, NCSN – Northern California Seismic Network, NOAA – National Oceanic and Atmospheric Administration, NTHMP – National Tsunami Hazards Mitigation Program, PGC – Pacific Geoscience Centre; PNSN – Pacific Northwest Seismic Network, PTWC – Richard H. Hagemeyer Pacific Tsunami Warning Center, USGS – U.S. Geological Survey, USNSN – U.S. National Seismic Network, WC/ATWC – West Coast/Alaska Tsunami Warning Center

1. Introduction

Before the widespread availability of high-speed computer networks, regional seismic networks in the U.S. operated as independent reporting entities. Seismic waveform data were, by necessity, telemetered to a single, regional center because of the high cost of long-distance communications. As a consequence, each center located earthquakes using only its own data. Although there were many seismic stations operating around the world, most of these data were not available to the tsunami warning centers in real time. Consequently, the warning centers had to make critical public safety decisions based on data recorded by their own seismic stations and a limited amount of continuous waveform data imported from the USGS U.S. National Seismic Network (USNSN). It could take as long as 5 minutes for seismic waves from earthquakes occurring in Cascadia (the coastal region of northern California to the Canadian border) to reach enough stations for the West Coast/Alaska Tsunami Warning Center (WC/ATWC) to accurately locate the earthquake and compute its magnitude. Since the first tsunami waves could reach the shore in tens of minutes, the added time due to seismic wave propagation made it even more difficult for the WC/ATWC to issue a warning in time to alert communities at risk.

In 1997, NOAA implemented the Tsunami Hazard Mitigation Implementation Plan (Tsunami Hazard Mitigation Federal/State Working Group, 1996) in cooperation with FEMA, USGS, and the five western states of Alaska, California, Hawaii, Oregon, and Washington. One aspect of the mitigation plan focused on improving the amount and quality of seismic data telemetered to the WC/ATWC and Richard H. Hagemeyer Pacific Tsunami Warning Center (PTWC). With funding provided by the National Tsunami Hazards Mitigation Program (NTHMP), the USGS was given the responsibility for upgrading seismic equipment and monitoring facilities of seismic networks operating in Cascadia, Alaska, and Hawaii. The goal of the project was to decrease the time required to issue a tsunami warning for earthquakes occurring within these regional networks and to reduce the likelihood that false tsunami warnings would be issued by providing seismic data to the tsunami warning centers.

2. Tsunami Warning Procedure

To understand how the NTHMP seismic project was designed to assist the tsunami warning centers in their assessment of the tsunamigenic potential of an earthquake, it is useful to briefly review how the centers operate. Time is of the essence in issuing a tsunami warning for earthquakes occurring offshore of the five western states, because the time interval between the origin of the earthquake and the time the first tsunami wave reaches land can be as little as 15 minutes. Even though automated monitoring systems are now capable of computing seismological parameters about an earthquake within minutes, the warning center seismologists must review this information to prevent the issuing of a false warning. Their response begins in less than 5 minutes after notification of the occurrence of a strong quake, and warning decisions must be made immediately and rapidly in the ensuing minutes. Consequently, the decision to issue a warning presently depends on only two criteria – magnitude (M) and location. Even if the main shock does not directly generate a tsunami through displacement of the seafloor, it is possible that secondary processes resulting from the quake could trigger submarine landslides or movement on secondary faults that could generate a tsunami. Thus, if the earthquake magnitude exceeds the specific warning criteria for a region (e.g., M 6.8 for Hawaii or M 7.0 for the west coast of the USA and Alaska) and locates offshore or near the coast, a warning is issued.

During this initial response there is little time to review other types of information such as depth, mechanism, and spectral content. Even if the warning center seismologists had access to this information during this time interval, the presentation of large volumes of information could conceivably slow down their response time and potentially confuse the analyst. Thus, the primary goal of the NTHMP seismic system is to provide the warning centers with more broadband, high-dynamic range seismic waveform data so that they can improve their ability to locate the earthquake and determine its magnitude.

After the tsunami warning centers issue an initial warning, they use additional data to either support their decision to continue or cancel the warning. This process is crucial because the decision to evacuate coastal communities at risk is disruptive and has large economic impacts on society. While tide-gauges and tsunameters (González *et al.*, this issue) provide the most important observations for this process, seismic data can provide additional information that can support this decision process. Thus, a secondary goal of the NTHMP seismic system was to rapidly provide the warning centers with related seismic information generated by other networks that could assist them in determining the tsunamigenic potential of earthquakes, such as ShakeMaps (Wald *et al.*, 1999) source mechanisms, distributions of small aftershocks, and earthquake source spectra.

3. Project Design

The NTHMP seismic project closely followed the preliminary seismic system design specifications provided in the Tsunami Hazard Mitigation Implementation Plan. The "Plan" called for (a) new instrumentation for seismic networks monitoring earthquakes in tsunamigenic regions in the U.S., (b) improved telemetry to the warning centers in order to enable the warning centers to rapidly receive the improved seismic information, and (c) rapid distribution of earthquake information to state emergency services agencies. The Plan called for upgraded instrumentation at approximately 36 seismic stations and the installation of 16 new stations in networks operated by the USGS in northern California and Hawaii and networks operated by universities in California, Oregon, Washington, and Alaska under support from the USGS. At all sites, the Plan called for the installation of digital dataloggers, broadband sensors to record waveforms from large teleseisms, and accelerometers to record on-scale waveforms from large, local earthquakes. Some of the identified stations already utilized digital dataloggers and had broadband instrumentation, but lacked accelerometers. About 30 sites utilized obsolete analog technology with only short-period sensors.

The Plan called for upgrading and replacing field telemetry links for regional seismic networks in northern California, and augmenting telemetry capacity in Oregon, Washington, Alaska, and Hawaii. Dedicated telemetry links from regional network processing centers to the PTWC and WC/ATWC were recommended to enable the warning centers to receive in real time the seismic data from the upgraded stations as well as from global stations recorded by the USNSN. Software was to be installed at all of the regional seismic networks to facilitate acquisition of the new seismic data and to build an appropriate interface so that the warning centers could receive this information. The goal was to enable the warning centers to be able to respond within 2 minutes after a major earthquake in the coastal regions of the five western states and have complete information on the earthquake within 5 minutes. With such rapid information, it would be possible for the warnings to be issued in advance of the first arrival of a tsunami wave.

Early in the design of the project it was recognized that seismic data from other networks performing global and regional earthquake monitoring could also be integrated into this expanded seismic network for the tsunami warning centers. In particular, dedicated telemetry links from the USNSN to the warning centers were proposed. Access to real-time waveform data from an increased number of global stations would enable the warning centers to respond more rapidly to teleseisms and reduce the likelihood of issuing false warnings. The NTHMP seismic system was also designed so that other seismological products generated by regional seismic networks could automatically be provided to the warning centers.

3.1. INSTRUMENTATION

Most of the instrumentation in use by regional seismic networks participating in the NTHMP seismic system was installed in the 1970s. A typical regional seismic station of this vintage consists of a single, vertical-component seismometer that continually transmits its data via analog telemetry to a central processing site. As a result of the limited dynamic range of the analog telemetry, the waveforms of most M > 2.0 earthquakes are clipped. Consequently, most regional networks routinely compute coda-duration magnitude, but this magnitude becomes increasingly unreliable above M 4.5. To increase the usefulness of regional network data to the warning centers, the data must have the extended frequency response and dynamic range to record the entire range of earthquake ground motion. To improve the dynamic range of the waveform data and ensure on-scale recording of all waveforms, the NTHMP seismic project installed dataloggers that generate 24-bit digital data at all stations.

No sensor currently has the capability to record the range of ground motion from teleseisms and local earthquakes from M 1.5 to greater than M 8. Broadband sensors provide the bandwidth to record long-period energy in waveforms generated by large earthquakes, and this makes it possible to compute accurate magnitudes, determine moment tensor solutions, and detect "slow" tsunami earthquakes (Kanamori and Kikuchi, 1993). However, their signal will clip in the near-field of local earthquakes as small as M 4.5. Accelerometers will remain on-scale during large, local earthquakes so that ShakeMaps can be generated, but they are too insensitive to record teleseismic waveforms.

To meet the above requirements, we installed tri-axial broadband sensors and accelerometers at 53 sites (Figure 1) using 24-bit digital dataloggers (Table I). At about half the sites we simply upgraded analog equipment, and for the remaining stations we installed equipment at new locations because nearby sites with existing analog equipment were unsuitable. At the beginning of the project, we performed evaluations of sensors and dataloggers and set minimum performance specifications. A variety of dataloggers (Güralp, Nanometrics, Reftek, and Quanterra) met these specifications and each network was allowed to purchase equipment most compatible with the local requirements of the installation[1]. Four different broadband sensors (Güralp

[1]Any use of trade names is for descriptive purposes only and does not imply endorsement by the USGS.

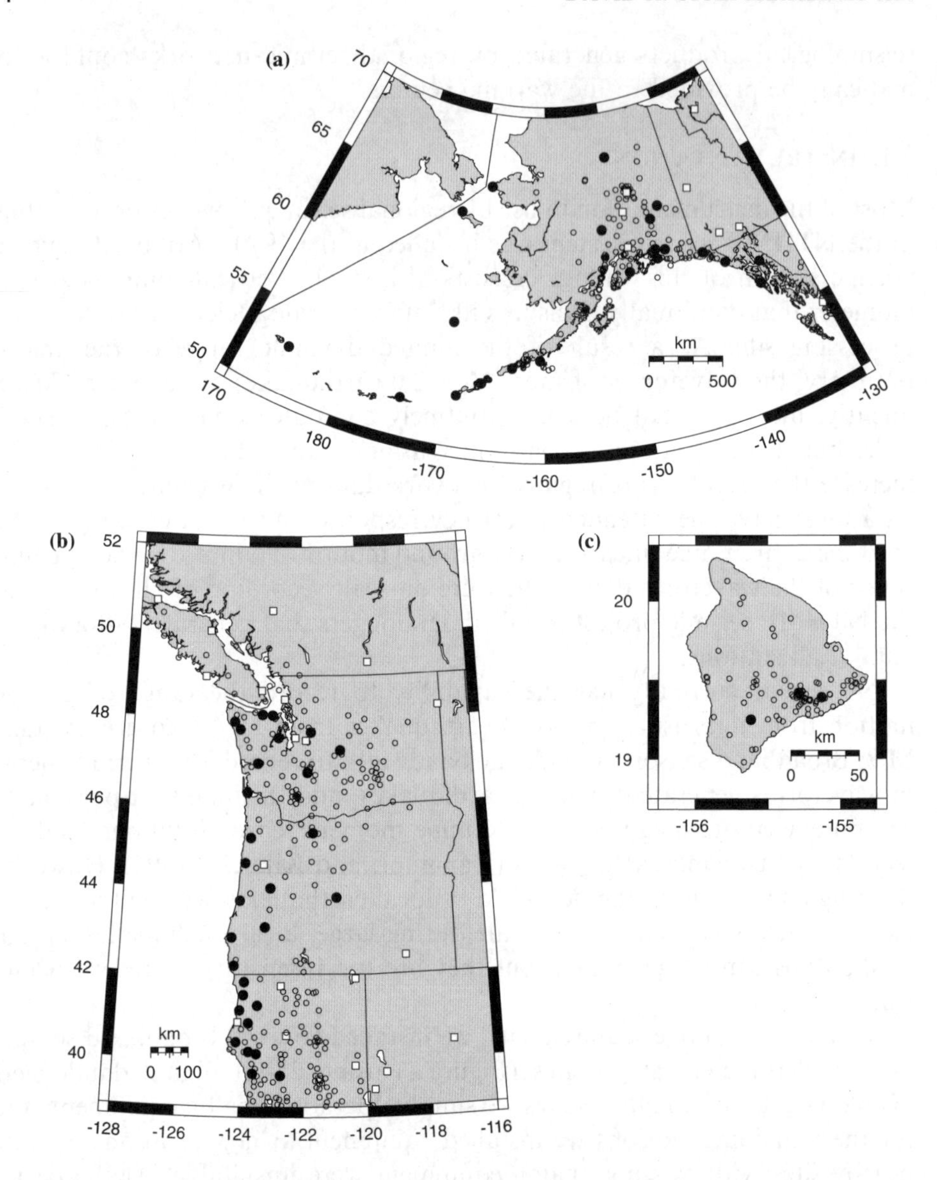

Figure 1. Map of seismic stations in (a) Alaska, (b) Cascadia, and (c) Hawaii. Solid circles indicate sites where new/upgraded seismic instrumentation was installed. Open circles indicate locations of existing, short-period, analog stations. White squares indicate locations of existing digital, broadband stations, such as those operated by the USNSN, University of California Berkeley, University of Oregon, University of Nevada Reno, and the Pacific Geoscience Centre in Vancouver.

CMG-40T, CMG-3T, and CMG-3ESP; and Streckheisen STS2) were purchased, but the same accelerometer (Kinematrics Episensor) was installed at all locations.

Table I. Seismic stations

No.	Network	Name	Location	Latitude	Longitude
1	AK	ATKA	Atka, AK	52.20	−174.20
2		BESS	Juneau, AK	58.30	−134.42
3		BMR	Bremner, AK	60.97	−144.60
4		COLD	Coldfoot, AK	67.25	−150.18
5		DCPH	Deception Hills, AK	59.07	−138.10
6		DIV	Chitina/Divide, AK	61.13	−145.77
7		DOT	Dot Lake, AK	63.65	−144.06
8		EYAK	Cordova, AK	60.55	−145.75
9		FALS	False Pass, AK	54.86	−163.42
10		GAMB	St. Lawrence Is., AK	63.78	−171.70
11		NIKO	Nikolski, AK	52.94	−168.87
12		PAX	Paxson, AK	62.97	−145.47
13		PIN	Pinnacle, AK	60.10	−140.26
14		PPLA	Purkeypile, AK	62.90	−152.19
15		SPIA	Saint Paul Island, AK	57.18	−170.25
16		SWD	Seward, AK	60.10	−149.45
17		TNA	Tin City, AK	65.56	−167.92
18		UNV	Unalaska, AK	53.85	−166.50
19	AT	SDPT	Sand Point, AK	55.35	−160.48
20		SIT	Sitka, AK	57.06	−135.32
21		SMY	Shemya, AK	52.73	−185.90
22	BK	GAS	Alder Springs, CA	39.65	−122.72
23	HV	KHU	Kahuku, HI	19.25	−155.62
24		STC	Steam Crack, HI	19.39	−155.13
25		UXL	Uwekahuna Vault, HI	19.42	−155.29
26	NC	KBO	Bosley Butte	42.21	−124.23
27		KCPB	Cahto Pk., CA	39.69	−123.58
28		KCT	CapeTown, CA	41.28	−123.45
29		KEB	Edson Butte, OR	42.87	−124.33
30		KHB	Hayfork Bally, CA	40.66	−123.22
31		KHMB	Horse Mt., CA	40.87	−123.73
32		KMPB	Mt. Pierce, CA	40.42	−124.12
33		KMR	Mail Ridge, CA	40.20	−123.71
34		KRMB	Red Mt., CA	41.52	−123.91
35		KRP	Rodgers, CA	41.16	−124.02
36		KSXB	Camp Six, CA	41.83	−123.88
37	UO	DBO	Dobson Buttes, OR	43.12	−123.24
38		PIN	Pine Mt., OR	43.81	−120.87
39		EUO	Eugene, OR	44.03	−123.07

Table 1. Continued

No.	Network	Name	Location	Latitude	Longitude
40	US	OCWA	Octopus Mt., WA	47.75	−124.18
41	UW	RWW	Ranney Well, WA	46.96	−123.54
42		GNW	Green Mt., WA	47.56	−122.83
43		LON	Longmire, WA	46.75	−121.81
44		SQM	PNNL – Sequim, WA	48.08	−123.05
45		LTY	Liberty, WA	47.26	−120.66
46		TAKO	Tahkenitch, OR	43.74	−124.08
47		MEGW	Megler, WA	46.27	−123.88
48		TTW	Tolt River, WA	47.69	−121.69
49		OFR	Forks, WA	47.93	−124.39
50		HEBO	Mt. Hebo, OR	45.21	−123.75
51		OPC	Port Angeles, WA	48.10	−123.41
52		TOLO	Toledo BPA, OR	44.62	−123.92
53		HOOD	Mt. Hood Meadows, OR	45.32	−121.65

AK University of Alaska Geophysical Institute, Alaska Earthquake Information Center; AT: NOAA/NWS West Coast/Alaska Tsunami Warning Center; BK: University of California Berkeley Seismological Laboratory; HV: USGS Hawaiian Volcano Observatory; NC: USGS Northern California Seismic Network; UO: University of Oregon Pacific Northwest Seismograph Network; US: USGS National Seismic Network; UW: University of Washington Pacific Northwest Seismograph Network.

3.2. COMMUNICATIONS AND REDUNDANCY

After several decades of monitoring, seismic networks have learned some painful lessons about telecommunications. Experience has shown that no form of communications is fail-safe. Power failures have brought down commercial telephone exchanges during large quakes. A satellite failure brought down large portions of the USNSN for weeks until remote satellite dishes were re-pointed. Telephone companies occasionally and unexpectedly take down Frame Relay networks for system upgrades. Operators of seismic networks learn from these situations and re-design their systems if possible. Despite this progress, there is no guarantee that a regional seismic network will continue to function when a great earthquake occurs in its region. Because tsunami warnings have life-safety implications, the NTHMP seismic project considered the impact on warning center operations in the event of a loss of critical seismic information in the epicentral area.

Because the cost and complexity of telemetry generally increase with distance, data from most seismic stations were telemetered to the closest regional network center. However, this situation is vulnerable to single points of failure if earthquake shaking disables a regional center or severs a critical communications link. To ensure that the system maintains the ability to report reliable information from the epicentral region, a small subset of stations in each network independently transmits seismic information to the USNSN via a satellite.

We designed a second level of redundancy into the network-to-network connectivity of the system. While the Internet offers essentially free telemetry between the regional network centers and the tsunami warning centers, it is not a reliable communication medium for applications with real-time reporting responsibilities because the bandwidth is not guaranteed and switches and routers may not operate during power outages. Accordingly,

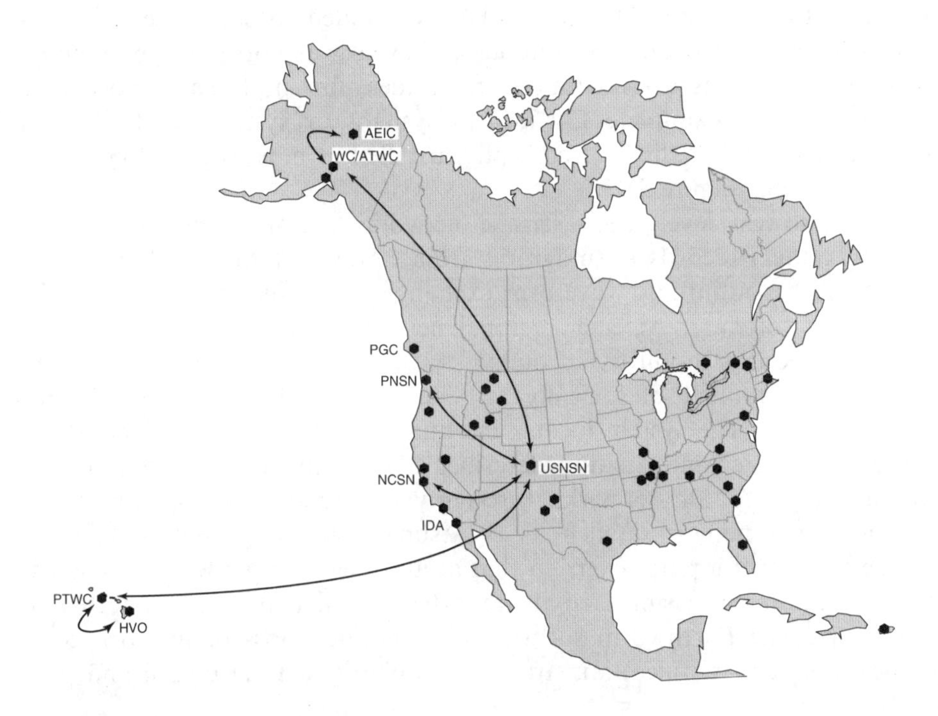

Figure 2. U.S. seismic networks (small octagons) capable of participating in real-time seismogram and parametric exchange utilizing Earthworm software. Solid lines indicate dedicated circuits connecting the Pacific Northwest Seismic Network (PNSN), the Northern California Seismic Network (NCSN), and the U.S. National Seismic Network (USNSN) with the Richard H. Hagemeyer Pacific Tsunami Warning Center (PTWC) and the West Coast/Alaska Tsunami Warning Center (WC/ATWC). Dedicated circuits also link the Alaska Earthquake Information Center (AEIC) to the WC/ATWC, and the Hawaiian Volcano Observatory (HVO) to PTWC. The PTWC and WC/ATWC also receive via the Internet continuous seismograms from other regional networks like the University of California Berkeley Seismological Laboratory, the California Institute of Technology Seismological Laboratory, and the Pacific Geoscience Centre (PGC), and from global seismic stations operated by institutions like the IDA network at Scripps Institution of Oceanography.

much of the data exchange between project participants is transmitted via dedicated, redundant commercial communication circuits (Figure 2). The utilization of satellite, dedicated point-to-point circuits, and Internet for telemetry to the tsunami warning centers provides some assurance against a single point-of-failure.

3.3. SEISMIC INFORMATION EXCHANGE

In 1993, the USGS began developing an earthquake reporting system for regional seismic networks called Earthworm (Johnson *et al.*, 1995). This system was initially designed to provide real-time earthquake reporting capability for regional networks. It later expanded to enable seismic processing centers to exchange continuous and event waveforms and parametric information such as arrival times, amplitudes, first-motions, hypocenters, and magnitudes. Complete Earthworm systems or systems that utilize a subset of its functionality are currently installed at most regional networks across the U.S. (Figure 2).

At the lowest level, each seismic network (regional, global, tsunami warning center) records seismograms from its own stations. All networks can establish continuous waveform exchange with other networks *via* the Earthworm software. At each regional network, the Earthworm software also continuously monitors incoming waveforms and declares an earthquake if the number of logically associated P-travel times exceeds some defined criteria. The system locates the quake, determines its magnitude as well as the related information described above, and makes this information available to all networks through established exchange protocols. Non-adjoining networks, such as the tsunami warning centers, can exchange hypocentral parameters so that at all times each network has access to information on seismic activity reported by all networks connected to the system. The Earthworm software also monitors loss of network communications, and notifications are immediately issued via e-mail and pagers.

At the initiation of the NTHMP seismic project, the software was not installed at the USGS Hawaiian Volcano Observatory (HVO), USNSN, PTWC, or WC/ATWC. At these networks, the project installed new computers, new multi-channel digitizers to make local, analog data accessible to the Earthworm system, and configured the systems in cooperation with local network operators. Later in the project, the Geological Survey of Canada Pacific Geoscience Centre in British Columbia installed Earthworm software so that they could participate in the project. The NTHMP project also developed a software interface to the IRIS/GSN system so that continuous data from these global stations could be sent to the tsunami warning centers.

4. Impact

The NTHMP seismic project was completed over a 5-year period and now provides the tsunami warning centers with real-time, high-dynamic range, broadband seismic data from regions of the western U.S., Alaska, and Hawaii where tsunamigenic earthquakes can occur, as well as from seismic stations around the world (Figure 3). These data decrease the time it takes to issue a tsunami alert by tens of minutes for earthquakes occurring outside the U.S. For earthquakes that occur within networks linked to the system, the warning time can be issued within a few minutes of the origin time. As a result of this project, the tsunami warning centers are able to issue more reliable and timely warnings to the public, decrease the likelihood of issuing false warnings, and decrease the likelihood of loss of life from tsunamis.

For example, before the NTHMP seismic project the PTWC only recorded eight short-period stations digitized at 20 samples/second (sps) and six long-period stations at 1 sps from outside of Hawaii using 12-bit resolution (McCreery, this issue). Now they obtain 20 and 40 sps data from about 90 broadband stations with 24-bit dataloggers. The time required to locate the earthquake is still governed by the time it takes for the P-waves to reach the most distant station in the network. Formerly, it would take the WC/ATWC 8–16 minutes depending on the location of the earthquake, but now the time has been shortened to 1–12 minutes. These added stations enable the warning centers to issue an alarm to the duty seismologists much sooner. More significant is that the time required to compute Ms on three stations has been

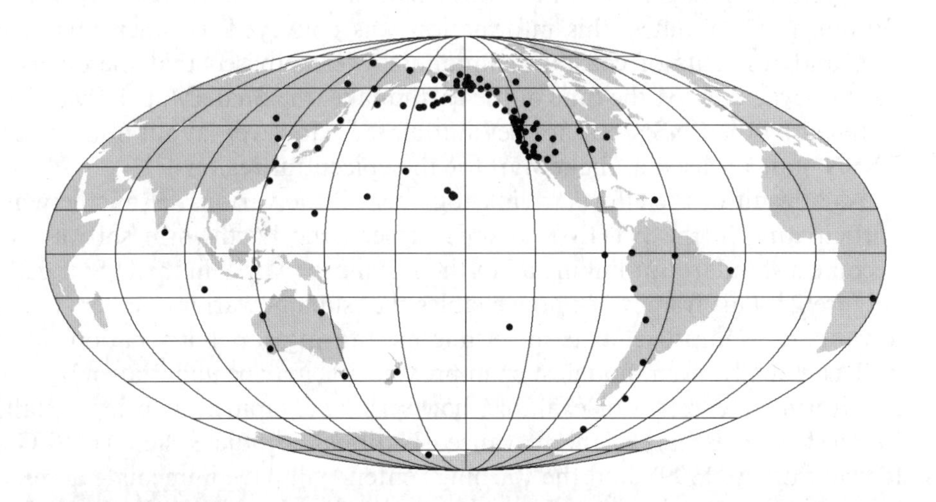

Figure 3. Global map showing distribution of 116 seismic stations providing continuous seismic waveforms through dedicated circuits installed for the NTHMP seismic project (Figure 2) and the Internet to the WC/ATWC. The PTWC also receives similar data from 90 stations throughout the world (McCreery, this issue).

greatly reduced. Whereas it formerly ranged from 5 to 55 minutes, it now takes a maximum of 20 minutes. In addition, with higher-dynamic range broadband data it is now possible for the warning centers to compute rapid estimates of the moment magnitude (Mw) from the initial P-wave (Tsuboi, 2000) instead of waiting for more slowly propagating surface waves.

The first opportunity to evaluate the response of the system for potentially tsunamigenic earthquakes in the U.S. occurred during the 28 February 2001 Mw 6.8 Nisqually earthquake beneath Olympia, Washington. The average response time of the WC/ATWC to issue a warning for the period 1982–2000 was 10.6 minutes (http://wcatwc.gov/wcatwc.htm). Because of new stations installed in the Pacific Northwest Seismic Network (PNSN) (Figure 1b), the WC/ATWC had access to waveform data in the epicentral area and their seismologists were able to locate the earthquake within 2 minutes after the origin time. In contrast, the automated software operated by the PNSN did not release its preliminary location until 5.5 minutes after the origin time. The initial WC/ATWC magnitude for the quake was 6.4 based on their observations of the initial body waves, and they issued a statement that the earthquake was not tsunamigenic based on both the location and magnitude of the earthquake. The final magnitude for the earthquake reported by the USGS National Earthquake Information Center was Mw 6.8 based on inversion of long-period body waves observed on data recorded globally, but this magnitude was not available for 1 hour 39 minutes after the earthquake. The WC/ATWC information about the Nisqually earthquake was automatically transmitted to Grays Harbor County, Washington via the NOAA Emergency Managers Weather Information Network (EMWIN) system. Within a few minutes, this information was conveyed to other emergency staff and residents of coastal communities were advised that there was no need to evacuate. At the time of the earthquake, the ShakeMap software was installed by the PNSN but not yet automated. However, about 6 hours later PNSN staff released a ShakeMap for the epicentral region.

As a result of the efforts to link together all the various seismic networks participating in the NTHMP seismic project, the Earthworm software has become a standard for linking all of the seismic networks in the U.S. together (Figure 2). This framework now enables the tsunami warning centers to take advantage of improvements in seismic monitoring capability nationwide as well as globally, even though such improvements may be undertaken by other monitoring agencies. For example, new seismic equipment is being installed by the USGS as part of the Advanced National Seismic System (U.S. Geological Survey, 1999), and the warning centers will have immediate access to this information via the system developed for the NTHMP. Data from seismic stations installed around the world by the other nations are now routinely exchanged in real time via the Internet, and the warning centers now have access to these data via this system. Moreover, as seismological institutions develop new algorithms such as ShakeMap (Wald *et al.*, 1999) and finite fault estimations (Dreger and Kaverina, 2000), they are incorporated into periodic releases of software. Since the warning centers utilize Earthworm software, they immediately have access to state-of-the-art methods for computing information about earthquakes.

5. Conclusion

The NTHMP seismic project has achieved the goals set forth in the Tsunami Hazard Mitigation Implementation Plan. The new seismic instrumentation deployed in the five western states has greatly improved the ability of the NOAA tsunami warning centers to respond to earthquakes in these regions, but the project has had an even greater impact than anticipated. The tsunami warning centers now have real-time access to an increased amount of seismic data collected around the world at no additional cost. Consequently, they are able to respond more rapidly to teleseismic earthquakes, compute more reliable locations and magnitudes, and be less likely to issue an erroneous tsunami warning.

The tsunami warning centers are now able to rapidly take advantage of seismological developments implemented by other seismic networks because they are using the Earthworm system. On the other hand, the NTHMP seismic project has served as a model for the rest of the regional seismic network community, and most of the networks supported by the USGS now utilize the Earthworm system to exchange seismic data. The synergism of all seismic networks and agencies participating in the NTHMP seismic project has advanced public safety by improving the capabilities of the tsunami warning centers and by improving the quality of information reported by regional seismic networks across the nation.

Acknowledgements

The success of the NTHMP seismic project was made possible by significant contributions from the following individuals: Tom Burdette, Jim Ramey, Pat Ryan, Dave Reneau, Dave Croker, Don Farrell, and Dan McNamara installed seismic stations. Pat Murphy, Rick Jensen, Charlene Fischer, Barbara Stocker, and Carol Lawson configured and installed the dedicated circuits for the project. Gray Jensen, Gary Holcomb, and Robert Uhrhammer performed instrument evaluations. Paul Okubo provided field support on the Big Island of Hawaii. Paul Whitmore, Alec Medbery, Guy Urban, Dave Ketchum, Kent Lindquist, and Stuart Weinstein provided software support. David Chavez wrote the software interface from the IDA/GSN network. We thank the staffs of the California Institute of Technology, University of California Berkeley, USGS National Seismic Network, and the IRIS Global Seismic Network for providing a real-time link to their seismic data. This manuscript was improved thanks to reviews by Jim Luetgert, Tom Yellin, Eric Geist, Eddie Bernard, and two anonymous reviewers.

References

Dreger, D. S. and Kaverina, A.: 2000, Seismic remote sensing for the earthquake source, process and near-source strong shaking: A case study of the October 16, 1999 Hector Mine earthquake. *Geophys. Res. Lett.* **27**, 1941–1944.

González, F. I., Bernard, E. N., Meinig, C., Eble, M. C., Mofjeld, H. O. and Stalin S.: 2005, The NTHMP Tsunameter Network. *Nat. Hazards* **35**, 25–39 (this issue).

Johnson, C. E., Bittenbinder, A., Bogaert, B., Deitz, L. and Kohler, W.: 1995, Earthworm: A flexible approach to seismic network processing. *Incorporated Research Institutions for Seismology (IRIS) Newsletter* **14**(2), 1–4.

Kanamori, H. and Kikuchi, M.: 1993, The 1992 Nicaragua earthquake; a slow tsunami earthquake associated with subducted sediments. *Nature* **361**, 714–716.

McCreery, C. S.: 2005, Impact of the National Tsunami Hazard Mitigation Program on operations of the Richard H. Hagemeyer Pacific Tsunami Warning Center. *Nat. Hazards* **35**, 73–88 (this issue).

Tsuboi, S.: 2000, Application of M_{wp} to tsunami earthquake. *Geophys. Res. Lett.* **27**, 3105–3108.

Tsunami Hazard Mitigation Federal/State Working Group.: 1996, Tsunami Hazard Mitigation Implementation Plan – A Report to the Senate Appropriations Committee. Appendices, 22 pp. http://www.pmel.noaa.gov/tsunami-hazard/hazard3.pdf.

U.S. Geological Survey.: 1999, Requirements for an Advanced National Seismic System. *U.S. Geological Survey Circular 1188*. 59 pp.

Wald, D. J., Quitoriano, V., Heaton, T. H., Kanamori, H., Scrivner, C. W. and Worden C. B.: 1999, TriNet "ShakeMaps": Rapid generation of instrumental ground motion and intensity maps for earthquakes in Southern California. *Earthquake Spectra* **15**, 537–556.

Natural Hazards (2005) 35: 73–88

Impact of the National Tsunami Hazard Mitigation Program on Operations of the Richard H. Hagemeyer Pacific Tsunami Warning Center

CHARLES S. McCREERY
NOAA, Richard H. Hagemeyer Pacific Tsunami Warning Center, 91-270 Ft. Weaver Rd., Ewa Beach, Hawaii, 96706, USA (Tel: +1-808-689-8207; Fax: +1-808-689-4543; E-mail: charles.mccreery@noaa.gov)

(Received: 22 January 2004; accepted: 27 April 2004)

Abstract. The first 7 years of the National Tsunami Hazard Mitigation Program (NTHMP) have had a significant positive impact on operations of the Richard H. Hagemeyer Pacific Tsunami Warning Center (PTWC). As a result of its seismic project, the amount and quality of real-time seismic data flowing into PTWC has increased dramatically, enabling more rapid, accurate, and detailed analyses of seismic events with tsunamigenic potential. Its tsunameter project is now providing real-time tsunameter data from seven strategic locations in the deep ocean to more accurately measure tsunami waves as they propagate from likely source regions toward shorelines at risk. These data have already been used operationally to help evaluate potential tsunami threats. A new type of tsunami run-up gauge has been deployed in Hawaii to more rapidly assess local tsunamis. Lastly, numerical modeling of tsunamis done with support from the NTHMP is beginning to provide tools for real-time tsunami forecasting that should reduce the incidence of unnecessary warnings and provide more accurate forecasts for destructive tsunamis.

Key words: tsunami, tsunami warnings, National Tsunami Hazard Mitigation Program, tsunami forecasts, tsunami models, Pacific Tsunami Warning Center, run-up detectors, tsunameter

Abbreviations: AFTAC – Air Force Technical Application Center, ASL – Albuquerque Seismological Laboratory, HVO – Hawaii Volcanoes Observatory, IDA – International Deployment of Accelerometers Project, LARC – Local Automatic Remote Collector, MOST – Method of Splitting Tsunamis, IASPEI – International Association of Seismology and Physics of the Earth's Interior, NEIC – National Earthquake Information Center, NOS – National Ocean Service, NTHMP – National Tsunami Hazard Mitigation Program, PTWC – Pacific Tsunami Warning Center, SHOA – Hydrographic Service of the Chilean Navy, SPLERT – Seismic Processing of Local Earthquakes in Real Time, USGS – US Geological Survey, WC/ATWC – West Coast/Alaska Tsunami Warning Center

1. Introduction

The United States operates two tsunami warning centers: the Richard H. Hagemeyer Pacific Tsunami Warning Center (PTWC) located in Ewa Beach,

Hawaii, and the West Coast/Alaska Tsunami Warning Center (WC/ATWC) located in Palmer, Alaska. WC/ATWC is responsible for local, regional, and distant tsunami warnings issued to Alaska, British Columbia, Washington, Oregon, and California. PTWC is responsible for local, regional, and distant tsunami warnings issued to Hawaii. It is also responsible for regional and distant tsunami warnings issued to American Samoa, Guam, and all other U.S. possessions and assets in the Pacific. In addition, as the operational center for the international Tsunami Warning System in the Pacific, PTWC issues warnings for regional and distant tsunamis in the Pacific Basin to almost every country around the Pacific rim and to most of the Pacific island states.

In general, the procedures used by PTWC to provide tsunami warnings are the following. Hardware and computer programs continually monitor seismic waveform data streams and alert watchstanders whenever large and widespread signals are detected from a significant earthquake. Watchstanders then locate the earthquake and determine its magnitude using a combination of automatic and interactive procedures. If the earthquake is shallow and is located under or very close to the sea, and if its magnitude exceeds a predetermined threshold, a warning is issued based on there being the potential that a destructive tsunami was generated. As sea level data are received from the nearest gauges, the tsunami can be confirmed if it exists and is measured. These measurements are then evaluated in the context of any historical events from the region, any applicable numerical simulations, and other predictive tools based on the earthquake and sea level parameters. Based on this evaluation the warning is continued, upgraded to cover a larger area, or cancelled. These procedures apply to the case of both a destructive teletsunami and a local or regional tsunami generated in Hawaii. The programs of the National Tsunami Hazard Mitigation Program (NTHMP) have enabled improvements in the speed, accuracy, and reliability of nearly all phases of this process.

2. Seismic Improvements – Teleseisms

As recently as 1996, PTWC relied on only a very limited set of seismic data to locate and determine the magnitude of distant earthquakes. Outside of Hawaii, the only continuous real-time waveform data received were from eight short-period and six low-gain long-period vertical seismometers located in Alaska and the continental U.S. These data were transmitted from the U.S. Geological Survey's (USGS) National Earthquake Information Center (NEIC) in Colorado to PTWC by modem over a dedicated circuit. The dynamic range of the data was very limited because the system was based on a 12-bit digitizer. In addition, the data were contaminated with frequent spikes from the 20-year-old hardware, so modern processing such as filtering or automatic arrival picking was not feasible. The data could, however, be used for event detection, manual arrival picking, and manual amplitude scaling for magnitude. Supplementing these data were time series data from Hawaii seismic stations (described below), automatic first arrival picks from NEIC, and first arrival times transmitted to PTWC from a few cooperating international observatories. These data were usually adequate for computing shallow epicenters to within a degree, but typically provided little depth control since the closest stations were often too far away to provide much constraint and depth phases were difficult to recognize on the narrow-band records. Lastly, computation of the surface wave magnitude, on which the warning criterion was based, was very slow for earthquakes in the southern or western Pacific due to the long delay as surface waves propagated to the U.S.

Beginning in about 1997, PTWC began importing data over the internet from a growing number of international broadband seismic stations having data available in near real time from data servers of the International Deployment of Accelerometers Project (IDA) and the USGS Albuquerque Seismological Laboratory (ASL). These data were generally of very high quality and gave PTWC new opportunities to improve its performance by utilizing their wider spatial coverage and by being able to apply more modern seismic analysis techniques to the tsunami warning problem.

Then in late 1999, the NTHMP's seismic project (Oppenheimer *et al.*, 2001) provided PTWC with hardware, software, communication circuits, and technical support for the USGS "Earthworm" seismic data exchange and processing system (Johnson *et al.*, 1995) to serve as a back end to PTWC's existing data processing environment. It allowed PTWC to receive continuous digital broadband seismic data from the U.S. National Seismic Network, from U.S. regional seismic networks that also operate "Earthworm" systems, and via NEIC from other worldwide networks such as the IRIS Global Seismic Network and Air Force Technical Application Center (AFTAC) Global Telemetered Seismic Network. It also allowed PTWC and WC/ATWC to more easily exchange their seismic and some of their sea level data continuously and in real time. In total, PTWC now receives data from about 90 broadband vertical seismic sensors located around the Pacific, including stations in S. America, Antarctica, New Zealand, Australia, SE Asia, Japan, Russia, and some Pacific islands, as well as in Alaska and the continental United States (Figure 1). These data are typically digitized at 20 samples per second with a 24-bit digitizer. These are the highest quality seismic data available with a wide dynamic range to stay on scale for all but the largest nearby earthquakes and with a frequency response that permits accurate timing of high-frequency P-wave arrivals at a few cycles-per-second and magnitude measurements at up to several hundred seconds period for the largest earthquakes. For reliability, PTWC operates two independent

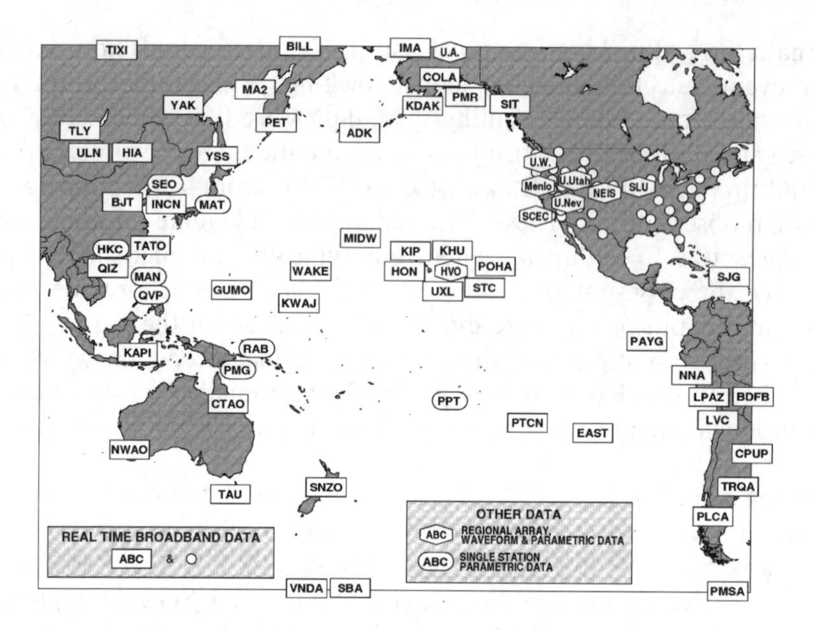

Figure 1. Seismic stations and networks providing waveform and parametric data to PTWC. The amount of data is more than an order of magnitude more data than was received just a few years ago.

"Earthworm" systems and uses both the dedicated NTHMP seismic project circuit and the public Internet to receive data. For additional reliability, PTWC continues to import data, about 40 broadband signals, directly over the internet from the IDA hub and from ASL that are outside the NTHMP seismic project network. While these data and systems are not flawless – there are sometimes extended station outages, data streams have intermittent gaps and overlaps, and whole systems and communication links occasionally fail – there is enough redundancy so sufficient data to accomplish PTWC's mission should be available except for under the most catastrophic circumstances. In such situations, PTWC and WC/ATWC can serve as backup centers for each other and NTHMP seismic project capabilities help facilitate this.

The new high-quality seismic data provide the foundation for improved warning center performance. Their extensive geographical distribution permits earlier detection of an earthquake and more rapid and accurate hypocenter calculations. The broader frequency band of the data and larger number of traces permit recognition of depth phases for more accurate depth determinations. In addition, many lower-frequency seismic waveforms allow techniques for determining moment magnitude, a more accurate measure of size than the surface wave magnitude for the largest earthquakes with the most tsunamigenic potential. PTWC now routinely calculates Mwp, the moment magnitude based on the first-arriving P waves (Tsuboi *et al.*, 1995), and mantle magnitude, Mm (Okal and Talandier, 1989), from surface waves that can be directly converted to moment magnitude. These computations are done for each of the available broadband seismic signals and final values are typically based on 30–50 independent measurements. As a result of having these capabilities, Mw was adopted in June 2003 as the magnitude to use in bulletins and for warning criteria. The new data also make possible automatic teleseismic epicenter determinations. PTWC has now implemented the teleseismic P-wave picker and associator developed by WC/ATWC and the two Centers exchange their automatic hypocenters as they are produced in the minutes following an earthquake. The broadband data also facilitate techniques for the discrimination of so-called "tsunami" or "slow" earthquakes that carry an especially high tsunamigenic potential (Kanamori, 1972). These events are usually recognized by unusually high ratios between low and high frequency seismic energy, and PTWC now routinely computes Mw-Ms and Theta values (Newman and Okal, 1998) as discriminants to check for this possibility. Additional analysis techniques are being developed, including rapid computation of the centroid moment tensor, slip distributions, and fault rupture dynamics. This type of source information is useful not only for quickly estimating tsunamigenic potential, but also to constrain initial conditions of the water wave numerical models used for forecasting tsunami impacts.

An important measure of how these seismic enhancements have helped improve PTWC performance is the elapsed time from the earthquake origin to bulletin issuance (Figure 2). From 1994 through 1998 it took 30–90 minutes to issue a bulletin. After 1999, when Earthworm and NTHMP

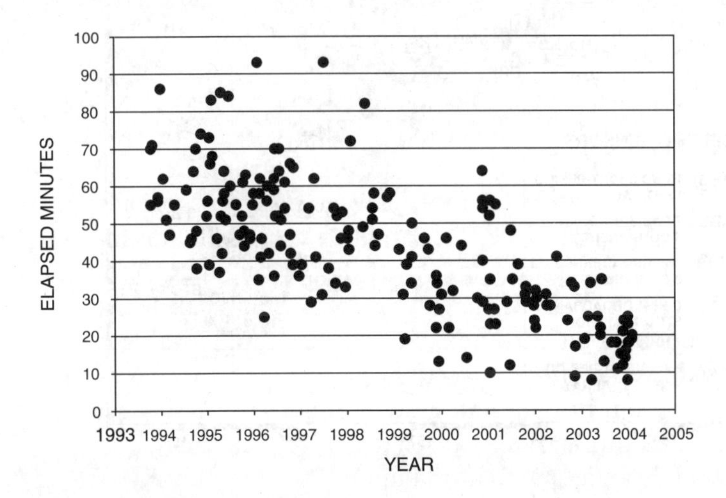

Figure 2. Elapsed minutes after a large Pacific earthquake to the issuance of a bulletin by PTWC containing a preliminary evaluation of the earthquake and its tsunamigenic potential.

seismic project circuits were installed, it took only 20–60 minutes. Since June of 2003, when procedures were officially changed to use Mw instead of Ms for magnitude criteria, it has taken just 25 minutes or less. This improved response time will help get warnings to areas at risk closer to the source where they are often needed most.

3. Seismic Improvements – Hawaiian Earthquakes

In its 200-year historical record, Hawaii has suffered two major local tsunamis, in 1868 and 1975. Both had maximum run-ups of around 15 m. A handful of smaller local tsunamis have also occurred. All historical events were generated on the volcanically and seismically active island of Hawaii, and their effects only felt on that island. However, models have shown a future tsunami could affect other islands in the chain. PTWC has the responsibility to provide warnings for such local and regional tsunamis, and while it may not be able to warn those nearest the epicentral region, its goal is to provide a warning to population centers more than just a few minutes away.

Prior to the NTHMP seismic project, PTWC only operated a regional array consisting of 10 low-gain short-period vertical seismometers, half distributed

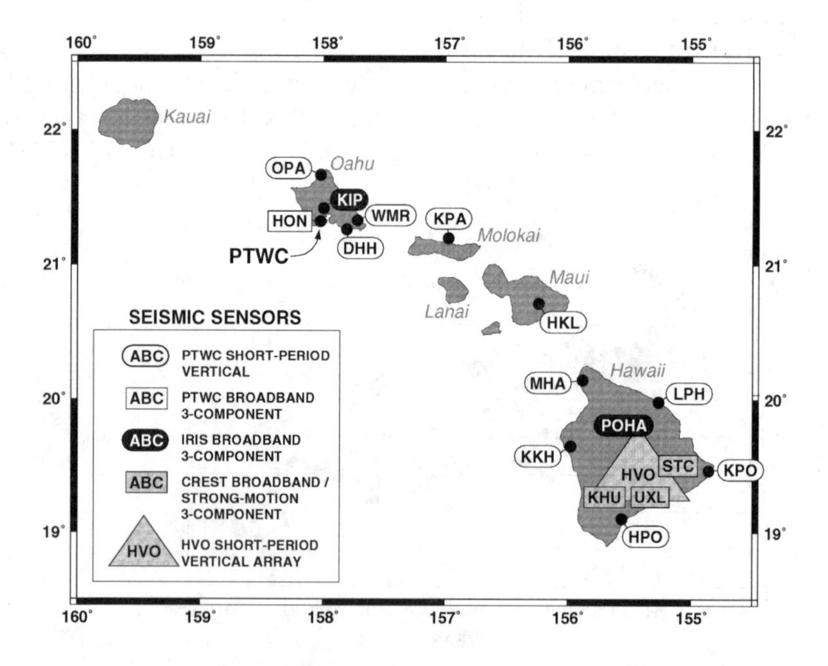

Figure 3. Seismic stations in the State of Hawaii used by PTWC for rapidly evaluating local earthquakes for their tsunamigenic potential. The most likely tsunami source regions are on the volcanically and seismically active island of Hawaii, particularly along its southeast and southwest facing coasts.

on the seismically active island of Hawaii, where all of Hawaii's historical local tsunamis have been generated, and the rest distributed further up the island chain (Figure 3). Supplementing these data were signals from eight high-gain vertical seismometers of the USGS's Hawaii Volcano Observatory (HVO), also located on Hawaii Island. In addition, PTWC operated the Honolulu (HON) station on Oahu, located just behind its operations center, with three-component short- and long-period seismometers. To locate an earthquake, PTWC relied on International Association of Seismology and Physics of the Earth's Interior (IASPEI) software to automatically detect and pick arrivals from the data streams and compute a hypocenter. This process was unreliable, however, and when it failed PTWC watchstanders manually picked arrivals and sent them to a location program – a very slow process when seconds count for a local tsunami warning. In 1997, PTWC implemented a special version of "Earthworm" called SPLERT (Seismic Processing of Local Earthquakes in Real Time), to receive automatic arrival picks made by HVO, associate them, and compute the hypocenter. After correcting a few flaws, this methodology has worked well and is now operational. It typically provides accurate hypocenters within 20–40 seconds of the earthquake origin.

The NTHMP seismic project has enhanced PTWC's Hawaiian earthquake capabilities in several ways. By installing a dedicated NTHMP seismic project circuit to HVO, PTWC now has more reliable access to the HVO arrival picks that were formerly sent only over the public Internet. PTWC is now also using a second dedicated circuit to HVO, funded by the Pacific Disaster Center, which adds an additional level of redundancy and reliability. The NTHMP seismic project also installed an "Earthworm" system at HVO to digitize and transmit more of its data streams to PTWC. At present, about 30 continuous short-period vertical signals are being transmitted, and these data are being automatically picked at PTWC with an "Earthworm" picker with arrival times forwarded to the aforementioned SPLERT regional associator-locator.

Lastly, the NTHMP seismic project installed three three-component broadband and strong-motion seismic stations on the island of Hawaii. These stations provide high-quality data that should stay on scale even for a major earthquake on that island. However, for a number of reasons they are not well suited to the local tsunami problem. Although the strong-motion accelerometers are likely to stay on scale for potentially tsunamigenic earthquakes on that island, their data require a problematic noise-introducing triple integration to generate the seismic moment time series from which Mwp is computed. The Richter local magnitude, ml, that could be computed more directly from the accelerometer data is unsuitable because it saturates in the mid-6 magnitude range where any significant tsunamigenic potential just begins. In addition, the NTHMP seismic project stations' proximity to likely sources means the long-period P-wave signal from those stations, on which Mwp is based, will be contaminated within seconds by the S-wave signal. This situation should be corrected by adding new stations or moving existing stations to islands further up the chain, particularly to Kauai, the furthest island. With significantly more spatial separation from likely sources, the broadband velocimeter data that only requires two integrations to get moment, is much more likely to stay on scale. In addition, there will be adequate separation between the P and S waves to properly measure Mwp. The Guralp CMG40T broadband sensors now being used at the NTHMP seismic project stations, with a response that falls off below 30-seconds period, should also be replaced in any new stations with broadband sensors having lower frequency sensitivity more suitable for measuring the very large earthquakes.

Again, the result of these efforts is clearly evidenced in the elapsed time for local tsunami bulletins (Figure 4). In 1996 it would take from 10 to sometimes more than 20 minutes to evaluate a local earthquake and issue a bulletin. After 1999 it took only 2–7 minutes and since mid-2002 only 2–4 minutes.

4. Sea Level Improvements – Pacific

Tsunami warnings are based initially only on seismic parameters. It is necessary to wait until a potential tsunami reaches the nearest sea level gauge to confirm or deny its existence and evaluate its character. Since the 1980s,

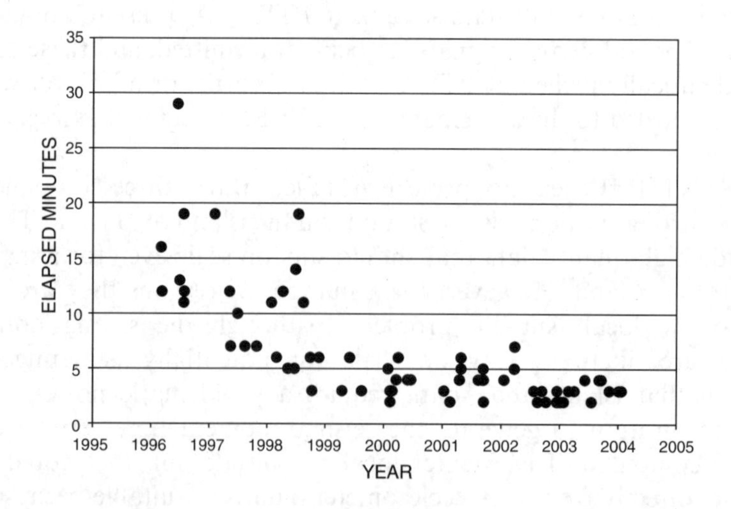

Figure 4. Elapsed minutes after a potentially felt Hawaii earthquake to the issuance of a bulletin by PTWC containing a preliminary evaluation of the earthquake and its tsunamigenic potential.

PTWC has received sea level data via satellite from stations around the Pacific for this purpose (Figure 5). The gauges currently number about a hundred and are operated by PTWC and various other organizations of the U.S., Japan, Russia, Chile, and Australia, often for a variety of purposes other than just tsunami detection and evaluation. While these data are much better than the Telex messages PTWC used to receive during an event from "tide observers" around the Pacific, they have significant shortcomings when being applied to the problem of tsunami forecasting. They are typically located in the shallow protected water of harbors and bays to provide security and a relatively benign ocean environment for instrument longevity. But in these environments tsunami waves coming in from the deep are highly modified in non-linear ways as they shoal and interact with the shoreline, severely limiting the predictive usefulness of the signals. In addition, since such gauges must be fixed to land (e.g., a pier), vast portions of the northern and eastern Pacific are not instrumented because there are no islands on which to site a gauge. Tsunamis from some of the most dangerous tsunamigenic zones stretching from northern Japan to Kamchatka to the Aleutian Islands and even down to Peru and Chile must go a long way before they reach the nearest strategically located gauge seaward of the source.

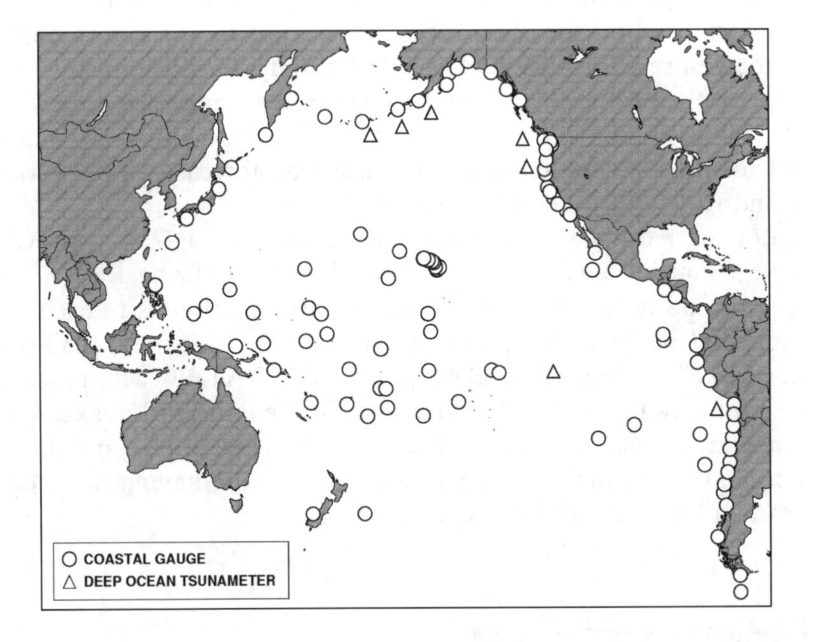

Figure 5. Sea level gauges used by PTWC to detect and evaluate tsunami waves propagating across the Pacific. Based on historical data, the most likely source regions for teletsunamis are the segments of the Pacific Rim that stretch from northern Japan to Kamchatka, from the western Aleutian Islands to the Gulf of Alaska, and along nearly the entire west coast of South America.

The tsunameter project (González *et al.*, this issue) addressed both of these shortcomings by developing a tsunameter – a deep ocean pressure gauge fixed to the ocean floor with satellite reporting through a nearby buoy. Capable of measuring sea level changes less than a centimeter, the tsunameter can detect tsunami waves and report them almost immediately to PTWC and WC/ATWC over an emergency satellite channel. Since the tsunameter can be sited in deep water, it can accurately record the character of tsunami waves as they propagate unaltered in the open ocean. In addition, the tsunameters can and have been sited strategically, directly between tsunamigenic zones and populated U.S. coastlines. At the time of this writing, seven tsunameters are in operation, with four more planned for deployment in the next 3 years. Three are off the Alaska Peninsula and Aleutian Islands and in a position to provide timely measurements of tsunami waves propagating toward Hawaii and the U.S. West Coast from tsunami sources in that region. Two more are off the coast of Washington and Oregon. They will provide timely measurements of tsunamis generated along the Cascadia subduction zone and also measure tsunami waves propagating toward Washington and Oregon from other areas of the Pacific. The sixth gauge is deployed just south of the equator in the eastern Pacific to provide readings of tsunamis generated in South America as they head toward Hawaii and the West Coast. A seventh gauge was recently deployed off the coast of northern Chile by the Hydrographic Service of the Chilean Navy (SHOA) to detect and measure tsunamis generated to the north of Chile. The ultimate utility of the tsunameter data won't be realized, however, until it can be interpreted using numerical tsunami simulations to provide comprehensive and accurate forecasts. This work is underway and is described below in Section 6.

In addition to receiving data from the tsunameters, PTWC and WC/ATWC have enhanced their sea level capabilities by utilizing the "Earthworm" systems and dedicated NTHMP seismic project circuits to exchange real-time regional sea level data. This provides PTWC with real-time data from eight Alaska/Aleutian sea level stations and WC/ATWC with real-time data from seven Hawaii stations. In the case of an Aleutian earthquake, when PTWC's tsunami evaluation must be made in an hour or less to give Hawaii Civil Defense adequate time to carry out an evacuation, having those data available in real time is extremely beneficial.

5. Sea Level Improvements – Hawaii

The State of Hawaii, with funds obtained through the NTHMP, contracted in 2001 for the purchase and installation of some newly developed remotely reporting tsunami run-up detectors (Figure 6). Eight were subsequently installed along Hawaii Island's southeast and southwest facing coasts – those

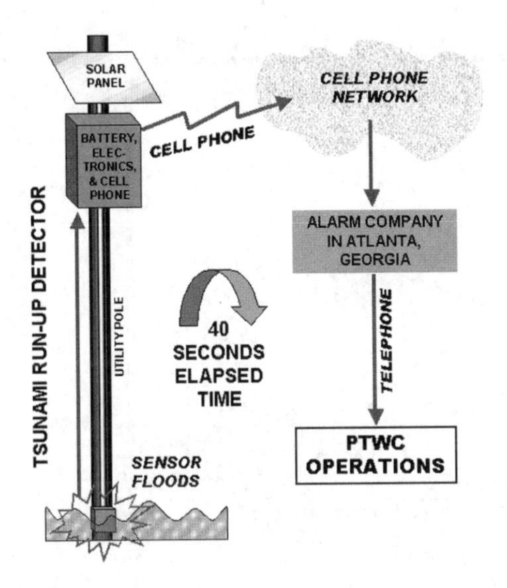

Figure 6. Tsunami run-up detectors used by PTWC are based on home security alarm technology. When the run-up sensor floods it sends a signal that is received by PTWC within about 40 seconds. The signal triggers alarms to immediately notify PTWC's duty personnel and the triggered sensor is displayed on a computer screen map.

most likely to be near the source of a local or regional tsunami (Figure 7). The run-up detectors will trigger and send a message back to PTWC within seconds of their sensor being flooded, positively indicating there is water on land. The sensors are 2.1–4.4 m above mean sea level and 18–119 m from the shoreline and are at locations where significant run-ups from past local tsunamis have been recorded. They are outside normal surf run-up and insensitive to rain or moisture other than a flood. Based on home security alarm technology and cell phone communications, the detectors are relatively inexpensive (about $1000 each), easy to install, and should be much easier to maintain than a normal sea level gauge. In addition, since the sensor does not have to be in the sea there are more options for siting them along the shore. In 2 years of operation there have been no false triggers, only one technical problem, and all detectors have triggered properly in manual bucket flooding tests conducted every 6 months.

The new run-up detectors are to help PTWC and the State with three key issues: (1) quickly determining if a tsunami on Hawaii Island is a threat to other islands in the chain, (2) improving the possibility of quickly detecting localized tsunamis generated by submarine landslides, and (3) reducing the chances of a false local warning. Based on recent numerical model results (Fryer *et al.*, 2001), a magnitude 7.2 earthquake on the southwest coast of Hawaii Island is capable of generating a statewide destructive tsunami. One characteristic of such a tsunami is that it would have large run-ups all along

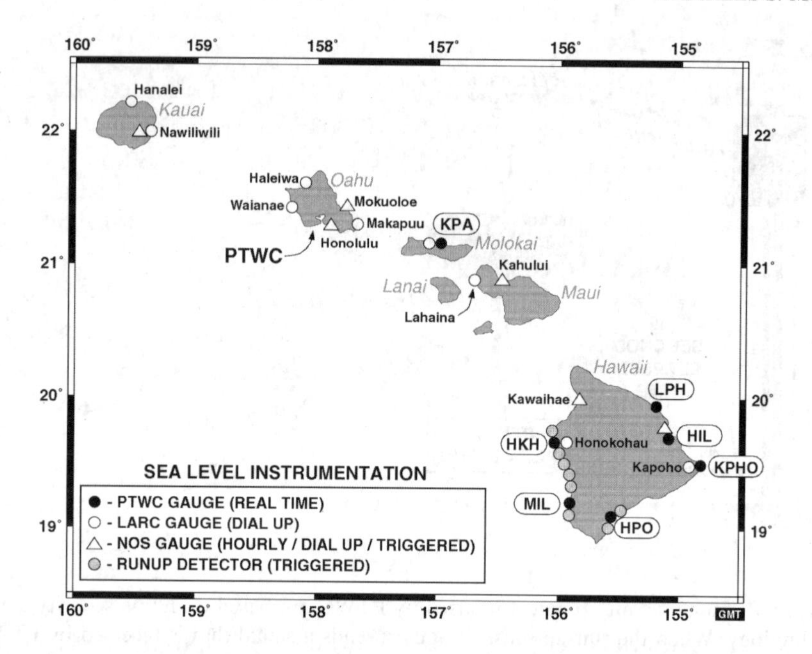

Figure 7. Sea level instruments used by PTWC to detect and evaluate distant or local tsunamis affecting the State of Hawaii. The PTWC gauges send their data continuously and in real time. The Local Automatic Remote Collector (LARC) gauges are dial-up only. National Ocean Service (NOS) gauges send their data via satellite in hourly transmissions, but will send more frequent data if they detect a tsunami. They can also be contacted by telephone. The run-up detectors signal almost immediately via cell phone if they have been flooded.

that southwest coast. Data from the run-up detectors will quickly indicate such widespread flooding and can therefore be used for making a timely decision regarding whether to urgently warn the rest of the State, and corresponding warning procedures are now in place at PTWC. Regarding the second issue, several smaller historical tsunamis have been generated in Hawaii by earthquakes below the magnitude threshold required for an automatic local tsunami warning. These are likely due to submarine landslides triggered by the earthquakes. Additional coverage provided by the run-up detectors means that in some cases such a tsunami might be detected in time to provide an effective warning. Warning procedures have also been developed for this scenario and are in place at PTWC. Regarding false local warnings, based on historical seismic data (Klein *et al.*, 2001), there are likely to be a few earthquakes each century that exceed PTWC's seismic criteria for an urgent local warning. These new run-up detectors should help PTWC more quickly confirm any significant local tsunami, and could eventually lead to a procedure whereby a warning is not issued unless an actual tsunami is confirmed.

6. Numerical Models for Forecasting

Since historical tsunamis are few and their data extremely limited, the only way to get detailed information about likely tsunami scenarios is to create synthetic data using numerical models. For the problem of real-time tsunami forecasting by a warning center, there is too little time to compute such synthetic data for a scenario as it is playing out, so precomputed model runs must be used. The precomputed synthetics that best fit the seismic parameters and sea level readings available at the time can form the basis for estimating impacts further afield. The success of this forecasting scheme depends upon several factors including: (1) the appropriateness and accuracy of the precomputed model runs, (2) the accuracy of the seismic parameters, (3) the availability and accuracy of sea level readings, and (4) the uniqueness of the fit. As described above, the NTHMP's seismic and tsunameter projects have enabled significant improvements for the seismic and sea level constraints on this problem.

A variety of numerical tsunami modeling has taken place in recent years that can be utilized by PTWC to implement this method of tsunami forecasting. This includes the MOST (Method Of Splitting Tsunami) model that was run for a comprehensive suite of Alaska-Aleutian sources (Titov and González, 1997; Titov *et al.*, 1999), the WC/ATWC model (Whitmore and Sokolowski, 1996) that has now been run for a variety of historical and hypothetical sources around the Pacific, and the Cheung model (Wei *et al.*, 2003), partially funded by the NTHMP through the State of Hawaii, that has been run for historical and hypothetical events in the Alaska–Aleutian region. For all of these models, synthetic deep sea records have been computed for comparison with tsunameter data as a key constraint for fitting an actual teletsunami scenario to a synthetic one. The NTHMP is now providing support for the development of an automated forecasting tool that will incorporate all three modeling approaches. The tool will ingest the seismic parameters and sea level data from coastal stations and tsunameters as it becomes available. Based on these constraints forecasts of near-shore tsunami waveforms, run-ups, and inundations at selected locations (Titov *et al.*, this issue) will be produced. The Centers will evaluate the forecasts based on the consistency between methods and on a variety of quality control factors and use the forecasts in their decision making.

As tsunami forecasting based on precomputed synthetics is still in its infancy it must be implemented cautiously. Nevertheless, it holds the promise of being the most useful tool available for rapid and accurate decision making by the tsunami warning centers, and for possibly basing regionalized and multi-level warnings in the future. For example, it may be possible to categorize run-up impacts as being <1 m, 1–3 m, or >3 m and devise corresponding responses appropriate to the threat. In addition, it may be possible to specify such levels for individual sections of coast. Such forecasting capabilities should make it possible to provide adequate warning protection to areas at risk while limiting the adverse impacts of full evacuations only to coasts where it is really necessary.

The usefulness of forecast modeling constrained by tsunameter data was illustrated on 17 November 2003 following a magnitude 7.5 earthquake in the Rat Islands. The event triggered a regional warning by WC/ATWC for the Aleutian Islands and an Advisory by PTWC for Hawaii. Based on the seismic parameters and historical data, a widespread destructive tsunami was not expected unless there were other contributing circumstances such as a large submarine landslide. The corresponding WC/ATWC pre-run model (Table I) for a shallow subduction earthquake of similar size and location also did not indicate destructive waves for Hawaii. As the scenario unfolded, the first sea level reading was from Shemya at 0.25 m amplitude, later increasing to 0.35 m. The next reading was from Adak with 0.06 m amplitude. Both readings agreed reasonably with the model, but the two Centers in their consultations agreed to wait until data from the tsunameter at 47°N, 171°W reading before issuing a cancellation. When the tsunameter data arrived about an hour after the earthquake with only 0.02 m amplitude, in agreement with the model, the warning was cancelled. Several hours later the tsunami swept across the Hawaiian island chain with amplitudes similar to what was predicted, a very encouraging result. The notable exceptions were at Haleiwa and Hilo where the measured tsunami amplitude was greater than the forecast by 120% and 55%, respectively.

Table I. Comparison of WC/ATWC model with gauge data for the 11/17/03 Rat Island tsunami

Location	Model amplitude* (m)	Gauge amplitude* (m)
Shemya, AK	0.31	0.35
Adak, AK	0.11	0.06
Tsunameter (47°N, 171°W)	0.02	0.02
Midway, HI	0.12	0.12
Hanalei, HI	0.17	0.18
Nawiliwili, HI	0.13	0.10
Haleiwa, HI	0.16	0.39
Mokuoloe, HI	0.08	0.01
Honolulu, HI	0.06	0.03
Kalaupapa, HI	0.13	0.19
Kahului, HI	0.22	0.22
Hilo, HI	0.11	0.17

* Measured center-to-crest, center-to-trough, or half crest-to-trough, with tide removed.

For the case of a local or regional Hawaii tsunami, Fryer *et al.*, (2001) have modeled scenarios for a variety of historical and hypothetical sources in Hawaii. This work has also been done with partial funding from the NTHMP through the State of Hawaii. It has been used as a foundation for recognizing statewide impacts from some local tsunami sources and for developing warning procedures that consider such events. At present, however, there are no plans for developing real-time local or regional forecasting capabilities.

7. Conclusions

The first 7 years of the NTHMP have enabled significant advances to the operational capabilities of PTWC. Largely as a result of the NTHMP seismic project, the Center now has real-time access to very high quality seismic data from stations around the Pacific and in Hawaii for more rapidly and comprehensively characterizing seismic sources that may trigger tsunamis. This has resulted in a significant reduction in elapsed time between the earthquake and PTWC's initial bulletin for both distant and local events. New deep-ocean sea level instrumentation has been developed and deployed to provide near real-time measurements of tsunami waves as they propagate unaltered in the mid-ocean toward threatened shorelines. These tsunameters permit more accurate tsunami assessments and fill gaps in coverage where, for example, there are no islands to site gauges. Inexpensive remotely reporting run-up detectors have also been developed for more rapidly confirming and assessing local tsunamis. Tools for tsunami forecasting based on numerical models are being developed to provide a better foundation for decision making to reduce false warnings and provide better predictions of tsunami severity. Based on the experience of the Rat Islands earthquake and tsunami of 17 November 2003, there is reason to be optimistic that such forecasting can work well and possibly lead to more region-specific and multi-level forecasts.

References

Fryer, G. J., Cheung, K. F., Smith, Jr., J. R., Teng, M. H., and Watts, P.: 2001, Inundation mapping in Hawaii. In: *Proceedings of the International Tsunami Symposium 2001 (ITS 2001)* (on CD-ROM), NTHMP Review Session, R-16, Seattle, WA, 7–10 August 2001, p. 207. http://www.pmel.noaa.gov/its2001/.

González, F. I., Bernard, E. N., Meinig, C., Eble, M. C., Mofjeld, H. O., and Stalin, S.: 2005, The NTHMP Tsunameter Network. *Nat. Hazards* **35**, 25–39 (this issue).

Johnson, C. E., Bittenbinder, A., Bogaert, B., Deitz, L., and Kohler, W.: 1995, Earthwork: a flexible approach to seismic network processing. *Incorporated Research Institutions for Seismology (IRIS) Newsletter,* **4**(2), 1–4.

Kanamori, H.: 1972, Mechanism of tsunami earthquakes. *Phys. Earth Planet. Inter.* **6**, 346–359.

Klein, F. W., Frankel, A. D., Mueller, C. S., Wesson, R. L., and Okubo, P. G.: 2001, Seismic hazard in Hawaii: high rate of large earthquakes and probabilistic ground-motion maps. *Bull. Seismol. Soc. Am.* **91**(3), 479–498.

Newman, A. V. and Okal, E. A.: 1998, Teleseismic estimates of radiated seismic energy: The E/M0 discriminant for tsunami earthquakes. *J. Geophys. Res.* **103**, 26885–26898.

Okal, E. A. and Talandier, J.: 1989, Mm: A variable-period magnitude. *J. Geophys. Res.* **94**(B4), 4169–4193.

Oppenheimer, D. H., Bittenbinder, A., Bogaert, B., Buland, R., Dietz, L., Hansen, R., Malone, S., McCreery, C. S., Sokolowski, T. J., and Weaver, C.: 2001, The CREST Project: Consolidated reporting of earthqukes and tsunamis. In: *Proceedings of the International Tsunami Symposium 2001 (ITS 2001)* (on CD-ROM), NTHMP Review Session, R-5, Seattle, WA, 7–10 August 2001, pp. 83–95. http://www.pmel.noaa.gov/its2001/.

Titov, V. and González, F. I.: 1997, Implementation and testing of the Method of Splitting Tsunami (MOST) model. Technical Report NOAA Tech. Memo. ERL PMEL-112 (PB98-122773), NOAA/Pacific Marine Environmental Laboratory, Seattle, WA.

Titov, V. V., González, F. I., Eble, M. C., Mofjeld, H. O., Newman, J. C., and Venturato, A. J.: 2005, Real-time tsunami forecasting: challenges and solutions. *Nat. Hazards* **35**, 41–58 (this issue).

Titov, V. V., Mofjeld, H. O., González, F. I., and Newman, J. C.: 1999, Offshore forecasting of Alaska-Aleutian Subduction Zone tsunamis in Hawaii. NOAA Tech. Memo. ERL PMEL-114, NOAA/Pacific Marine Environmental Laboratory, Seattle, WA.

Tsuboi, S., Abe, K., Takano, K., and Yamananka, Y.: 1995, Rapid determination of Mw from broadband P waveforms. *Bull. Seismol. Soc. Am.* **85**, 606–613.

Wei, Y., Cheung, K. F., Curtis, G. D., and McCreery, C. S.: 2003, Inverse algorithm for tsunami forecasts. *J. Waterw. Port Coast. Ocean Eng.* **129**(3), 60–69.

Whitmore, P. M. and Sokolowski, T. J.: 1996, Predicting tsunami amplitudes along the North American coast from tsunamis generated in the northwest Pacific during tsunami warnings. *Sci. Tsunami Haz.* **14**, 147–166.

Natural Hazards (2005) 35: 89–110

Progress in NTHMP Hazard Assessment

FRANK I. GONZÁLEZ[1*], VASILY V. TITOV[2], HAROLD O. MOFJELD[1], ANGIE J. VENTURATO[2], R. SCOTT SIMMONS[3], ROGER HANSEN[4], RODNEY COMBELLICK[5], RICHARD K. EISNER[6], DON F. HOIRUP[7], BRIAN S. YANAGI[8], STERLING YONG[9], MARK DARIENZO[10], GEORGE R. PRIEST[11], GEORGE L. CRAWFORD[12] and TIMOTHY J. WALSH[13]

[1]NOAA/Pacific Marine Environmental Laboratory, Seattle, WA 98115, USA; [2]University of Washington, Joint Institute for the Study of the Atmosphere and Ocean (JISAO), Seattle, WA 98195-4235, USA; [3]Alaska Division of Emergency Services; [4]University of Alaska Geophysical Institute and Office of the State Seismologist; [5]Alaska Department of Natural Resources, Division of Geological & Geophysical Surveys; [6]CISN and Earthquake Programs, Governor's Office of Emergency Services, Oakland, CA 94610-2421, USA; [7]California Geological Survey; [8]Hawaii Civil Defense Division, Honolulu, HI, USA; [9]Hawaii Department of Land and Natural Resources, Engineering Division; [10]Oregon Emergency Management, Salem, OR 97309-5062, USA; [11]Oregon Department of Geology and Mineral Industries, Coastal Field Office, Newport, OR 97365 USA; [12]Washington State Military Department, Emergency Management Division, Camp Murray, WA 98430-5211, USA; [13]Washington Department of Natural Resources, Division of Geology and Earth Resources, Olympia, WA 98504-7007, USA

(Received: 4 February 2004; accepted: 27 April 2004)

Abstract. The Hazard Assessment component of the U.S. National Tsunami Hazard Mitigation Program has completed 22 modeling efforts covering 113 coastal communities with an estimated population of 1.2 million residents that are at risk. Twenty-three evacuation maps have also been completed. Important improvements in organizational structure have been made with the addition of two State geotechnical agency representatives to Steering Group membership, and progress has been made on other improvements suggested by program reviewers.

Key words: tsunami modeling, mapping, hazard assessment, National Tsunami Hazard Mitigation Program, emergency management, evacuation

Abbreviations: ADES – Alaska Division of Emergency Service, DOGAMI – Oregon Department of Geology and Mineral Industries, DNR – Department of Natural Resources, EM – Emergency Management, FEMA – Federal Emergency Management Agency, NEES – Network for Earthquake Engineering Simulation, NOAA – National Oceanic and Atmospheric Administration, NOS – National Ocean Service, NSF – National Science Foundation, NTHMP – National Tsunami Hazard Mitigation Program, OEM – Oregon Emergency Management, OGI – Oregon Graduate Institute, TAC – Technical Advisory Committee,

*Author for correspondence: Tel.: +1-206-526-6803; Fax: +1-206-526-6485; E-mail: frank.i.gonzalez@noaa.gov

TIME – Tsunami Inundation Mapping Efforts, UAGI – University of Alaska Geophysical Institute, USGS – U.S. Geological Survey

1. Background

The U.S. National Tsunami Hazard Mitigation Program (NTHMP) was established in 1997 as a partnership of the five Pacific States of Alaska, California, Hawaii, Oregon, and Washington with four Federal Agencies – the National Oceanic and Atmospheric Administration (NOAA), the U.S. Geological Survey (USGS), the Federal Emergency Management Agency (FEMA), and the National Science Foundation (NSF). Led by NOAA, the overriding goal of the NTHMP is mitigation of the tsunami hazard to all threatened U.S. coastal communities (Tsunami Hazard Mitigation Federal/State Working Group, 1996). Site-specific Hazard Assessment is an important component of achieving this goal and, from the start, there has been continuing, unanimous recognition and agreement among NTHMP partners that inundation and evacuation maps are the fundamental basis of local tsunami hazard planning. Without a clear understanding of what areas are at risk and which areas are unlikely to be flooded, it is impossible to develop effective emergency response plans, educational programs, and outreach efforts.

In response to this need, the NTHMP Hazard Assessment component initiated the first systematic, national effort to transfer state-of-the-art tsunami inundation modeling technology from the research environment to an operational setting for routine production of inundation maps. Numerical models have therefore been a primary tsunami Hazard Assessment tool since the inception of the NTHMP. *Scientific* products based on numerical simulations – including, e.g., maximum inundation and maximum current speed estimates (Figure 1), animations, and GIS data files – provide State and local officials with a basic assessment of the site-specific hazard posed by potential tsunami events. *Emergency management* products and activities – including, e.g., State-produced inundation maps (Figure 2) and evacuation maps (Figure 3), brochures, community meetings, workshops, and other educational and outreach efforts – are then developed with the aid of the modeling products and historical and other data. This process supports the missions of Federal agency partners and assists State agencies in meeting responsibilities that are, in most cases, explicitly imposed by legislation.

The long-term goals of the NTHMP Hazard Assessment component are to

1. Develop inundation maps and associated scientific and emergency management products for every U.S. coastal community at risk
2. Import and implement continuing advances in modeling and emergency management technology and methods

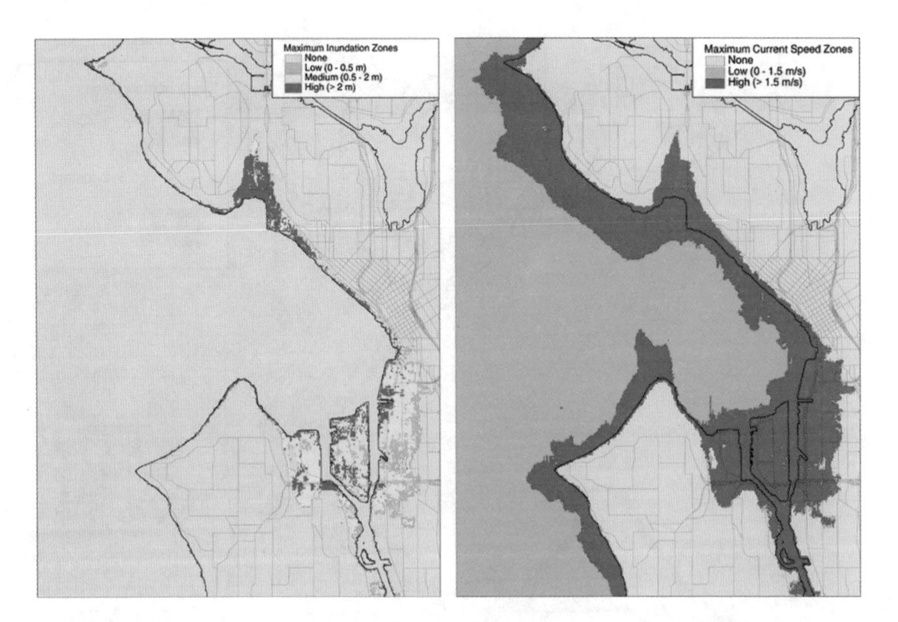

Figure 1. Tsunami inundation modeling products for Seattle, Washington (from Titov *et al.*, 2003). Left panel: zoned estimates of maximum inundation depth. Right panel: zoned estimates of maximum currents.

3. Establish a systematic program to review and upgrade the existing scientific and emergency management products

Initially, relatively simple one-dimensional (1-D) modeling technology was considered, because this approach appeared to require fewer resources than the more advanced two-dimensional (2-D) technology now available. However, on comparison of these two technologies, the decision was made to utilize 2-D modeling technology for all mapping. It was recognized that adoption of 2-D modeling technology would reduce the pace of modeling and mapping, but that the result would be products of indisputably improved detail, quality, and reliability. To optimize effective use of the limited time and resources available, NTHMP partners have agreed to assumption of the following responsibilities:

- State Agencies identify the high-priority communities to be mapped.
- Tsunami modeling scientists utilize 2-D models to provide State Agencies with inundation modeling products for high-priority areas.
- State Agencies and local officials produce and publish official inundation and evacuation maps, using inundation modeling products as guidance.
- The NOAA Center for Tsunami Inundation Mapping Efforts (TIME) assists State Agencies and tsunami modelers with the modeling and mapping effort.

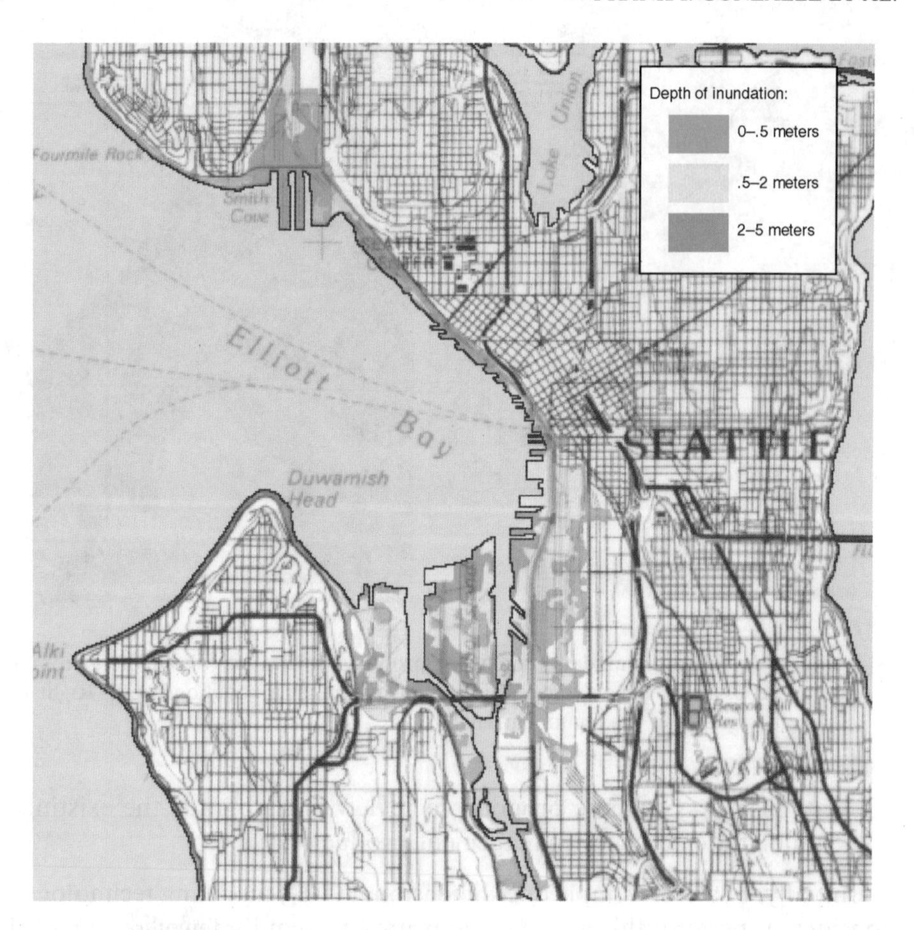

Figure 2. Tsunami inundation map for Seattle, Washington (from Walsh *et al.*, 2003b) produced and published by Washington State, using modeling products as guidance.

The pursuit of ambitious goals with finite resources demands effective program philosophies and strategies, and the continuing success of the NTHMP is due in part to the careful development of an initial implementation plan (Tsunami Hazard Mitigation Federal/State Working Group, 1996). A successful enterprise must also continually evolve and benefit from lessons learned. The NTHMP is no exception, and a Steering Group monitors performance and guides this evolution. Hazard Assessment accomplishments for the period 1997–2001 were discussed during a formal review of the NTHMP on 7 August 2001 and documented in a set of reports: Bernard (2001), Crawford (2001), Eisner *et al.* (2001), Fryer *et al.* (2001), González *et al.* (2001), Hansen *et al.* (2001), Priest *et al.* (2001). This article summarizes Hazard Assessment accomplishments to date, reviews the evolving implementation strategy, discusses continuing scientific and technical issues,

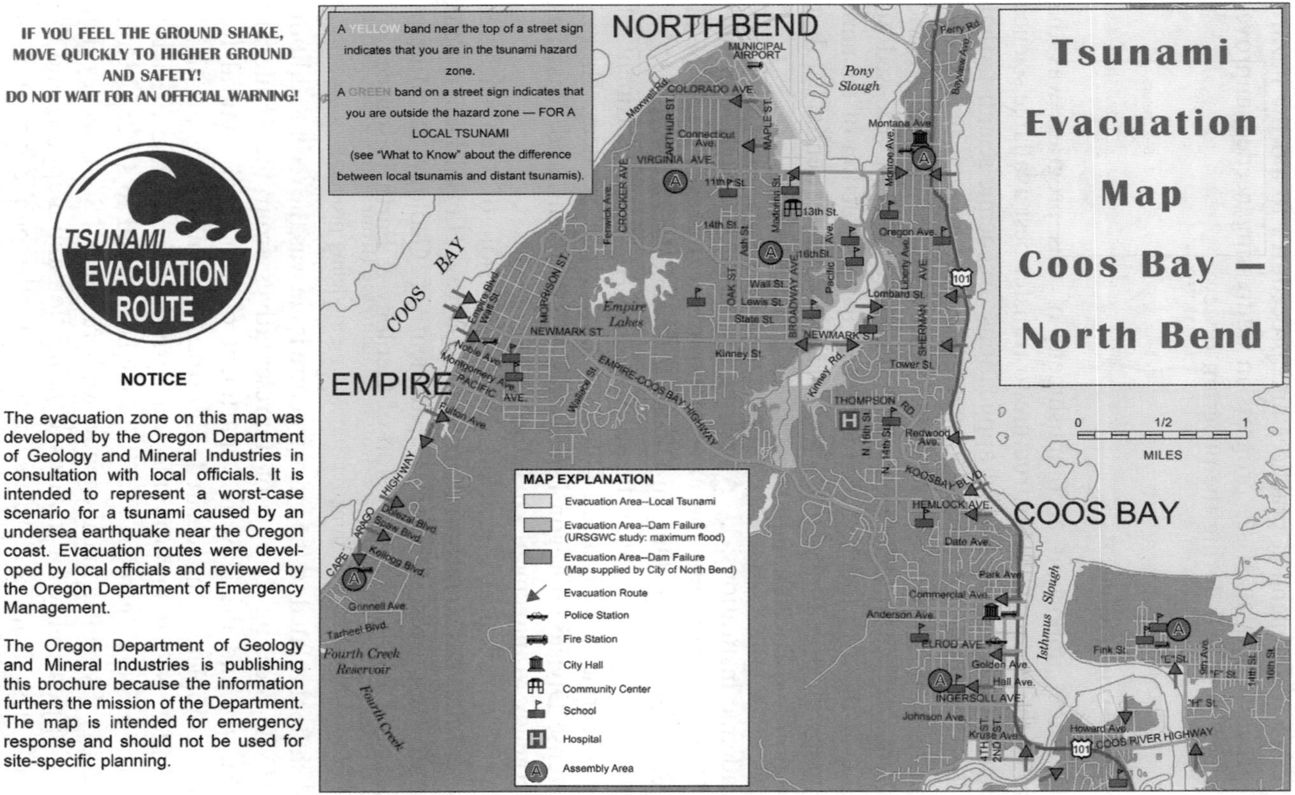
IF YOU FEEL THE GROUND SHAKE, MOVE QUICKLY TO HIGHER GROUND AND SAFETY!
DO NOT WAIT FOR AN OFFICIAL WARNING!
TSUNAMI EVACUATION ROUTE
NOTICE
The evacuation zone on this map was developed by the Oregon Department of Geology and Mineral Industries in consultation with local officials. It is intended to represent a worst-case scenario for a tsunami caused by an undersea earthquake near the Oregon coast. Evacuation routes were developed by local officials and reviewed by the Oregon Department of Emergency Management.
The Oregon Department of Geology and Mineral Industries is publishing this brochure because the information furthers the mission of the Department. The map is intended for emergency response and should not be used for site-specific planning.
A YELLOW band near the top of a street sign indicates that you are in the tsunami hazard zone.
A GREEN band on a street sign indicates that you are outside the hazard zone — FOR A LOCAL TSUNAMI
(see "What to Know" about the difference between local tsunamis and distant tsunamis).
Tsunami Evacuation Map
Coos Bay — North Bend
NORTH BEND
EMPIRE
COOS BAY
MUNICIPAL AIRPORT
Pony Slough
Empire Lakes
Isthmus Slough
Fourth Creek Reservoir
Fourth Creek
0 1/2 1
MILES
MAP EXPLANATION
Evacuation Area--Local Tsunami
Evacuation Area--Dam Failure (URSGWC study: maximum flood)
Evacuation Area--Dam Failure (Map supplied by City of North Bend)
Evacuation Route
Police Station
Fire Station
City Hall
Community Center
School
Hospital
Assembly Area

Figure 3. Tsunami evacuation map for the Coos Bay – North Bend, Oregon area (PDF available at http://www.oregongeology.com/earthquakes/Coastal/Tsubrochures.htm).

presents a set of recommended practices and procedures to aid the hazard assessment process, and discusses progress in addressing needed improvements that were identified during the program review.

2. Progress and Accomplishments

Arguably the most important accomplishment of this program was the development of the necessary infrastructure to transfer best available science from research settings to operational applications. Academic scientists that were given the opportunity to apply their tsunami modeling expertise to real-world problems responded to this challenge with enthusiasm. Prior to the NTHMP, there were no modeling groups that conducted R&D focused on applying state-of-the-art modeling technology to the production of tsunami inundation maps for operational use. The NTHMP was the essential catalyst for the initiation and coordination of such R&D activities at

- Four academic institutions: U. Alaska at Fairbanks, U. Southern California, U. of Hawaii, and the Oregon Graduate Institute's School of Science & Engineering at Oregon Health & Science University.
- Five State Emergency Management Agencies: Alaska Department of Emergency Services, California Office of Emergency Services, Hawaii Civil Defense Division, Oregon Emergency Management, and Washington Emergency Management Division.
- Five State Geotechnical Agencies: Alaska Division of Geological & Geophysical Surveys, California Geological Survey, Hawaii Department of Natural Resources, Oregon Department of Geology & Mineral Industries, and Washington Division of Geology & Earth Resources.
- The NOAA TIME Center.

An accomplishment related to the infrastructure issue was the recent addition to the NTHMP Steering Group of representatives from the California Geological Survey and the Hawaii Department of Land and Natural Resources. This completed the desired Hazard Assessment organizational structure and addressed an important lesson learned early on – that the active participation of a State "geotechnical" agency is essential to success (González *et al.*, 2001). By geotechnical agency, we here mean a State agency that bears primary responsibility for hazards identification and mapping, and the distribution of these maps and related reports to the public, i.e., a State counterpart to the U.S. Geological Survey.

Table I summarizes the progress of the NTHMP Hazard Assessment effort to date. A bibliography of reports related to the work summarized in this table is available on the TIME Center "Resources" web page (http://www.pmel.noaa.gov/tsunami/time/resources/). The entries for at-risk popu-

Table I. Summary of modeling and mapping progress to date. Only 2-D modeling efforts completed as part of the U.S. NTHMP Hazard Assessment effort are listed. Not included, for example, are Oregon maps developed before the NTHMP effort. Similarly, Hawaii had already provided all at-risk communities with 66 evacuation maps that were developed with 1-D models before the NTHMP effort, and has only recently initiated a 2-D modeling program to update these maps. Here, a "modeling effort" refers to the process described in section 4.3. Each effort employs a computational grid system that provides coverage at a horizontal spatial resolution of 50 m or less for one or more communities, and coverage with coarser resolution in the remaining areas of the model computational region, which varies in size but is typically on the order of 100 km × 50 km. Estimates of the at-risk population covered by the modeling efforts were obtained with the first version of a GIS-based algorithm under continuing improvement (González *et al.*, 2001). Evacuation maps are subsequently developed with the guidance of modeling products.

State	Modeling Efforts	Communities Covered	At-risk Population	Evacuation Maps
Alaska	2	5	9608	4
California	7	58	1,074,426	0
Oregon	8	25	62,894	17
Washington	5	25	44,383	2
Total	22	113	1,191,311	23

lation covered by the modeling efforts were computed by the TIME Center with a GIS-based algorithm that identifies Census 2000 blocks within a kilometer of the coast (González *et al.*, 2001). Development of the algorithm continues, and work is underway to include census block vertical elevation as an at-risk criterion and replace the coastal distance criterion with computed maximum inundation lines.

Interpretation of Table I requires an understanding of differences among the five States that affect progress in hazard assessment. These differences include physical characteristics, coastal population density and development, history of tsunami events, previous hazard assessment efforts, the scientific and technical infrastructure, the availability of adequate bathymetric and topographic data, the level of knowledge regarding potential tsunami sources, and the State organizational structure. A discussion of these issues and an overview of NTHMP Hazard Assessment efforts, including a summary of accomplishments during the first 5 years of the program, are provided by González *et al.* (2001). The subsections that follow provide a brief review of Hazard Assessment activities in each State, including a short history of the effort, progress to date, ongoing projects, and plans.

2.1. ALASKA

Funding for Hazard Assessment was first allocated in 1998 to the Alaska Division of Emergency Services (ADES). In Alaska, geotechnical guidance is provided by the University of Alaska Geophysical Institute (UAGI), the Office of the State Seismologist, and the Division of Geological and Geophysical Surveys of the Alaska Department of Natural Resources. The latter agency is also responsible for map publication and distribution services. Prior to receipt of funding, a significant effort was initiated to develop the necessary infrastructure at the UAGI for tsunami inundation modeling. Concurrently, the ADES organized a careful, systematic analysis of Alaskan coastal communities (Table II). Three on Kodiak Island were selected for the first inundation modeling effort – Kodiak City, Women's Bay, and the U.S. Coast Guard Base – and a prioritized list was developed for the next nine communities to be modeled (Suleimani *et al.*, 2002a, b). Source specification involves development of a "credible worst case" scenario for a particular site, including simulation of the 1964 Prince William Sound earthquake and tsunami that devastated the area.

Alaska presents a significant Hazard Assessment challenge. It has the longest coastline and the lowest population density, but a great many communities at risk. Of the five States, it is the most difficult in which to assemble bathymetry and topography data suitable for inundation modeling and, from the outset, this issue has forced deviations in the sequence of modeling efforts from original State priorities. The fourth and fifth priority communities of Homer and Seldovia were mapped before three higher priority communities – Seward, Sitka, and Sand Point – that lacked adequate bathymetric and topographic data. The Hazard Assessment team has communicated Alaska data priorities to NOAA's National Ocean Service and the USGS, which are now including these priorities as a factor in their project planning.

Since the Kodiak effort was completed in 2001, the Homer-Seldovia inundation map has been completed and four evacuation maps have been produced. Modeling of Seward is underway, and computational grids are being developed for the Sitka and Yakutat areas. Existing maps and reports can be viewed and downloaded from the Division of Geological and Geophysical Surveys website: http://www.dggs.dnr.state.ak.us/.

2.2. CALIFORNIA

The highest coastal population density and development are found in California, where the Office of Emergency Services directs Hazard Assessment efforts that were initiated in 1998 with inundation modeling conducted by the University of Southern California. In 2002, the California Geological Survey agreed to provide geotechnical guidance, map production and distribution services, and a scientist to serve on the NTHMP Steering Group.

California hazard assessment priorities are driven primarily by consideration of the population and infrastructure that are at risk. The development of "credible worst case" sources is conducted in consultation with experts on local and regional seismic fault systems and landslide potential. Since 2001, two additional inundation modeling efforts have been completed. Because of the high population density, the current California total of seven mapping efforts covers 58 communities. Completed inundation maps are distributed to individual counties, which use them as guidance to develop evacuation maps. Work is now underway for Orange County, Ventura County, San Luis Obispo County, and the San Francisco Bay area. A web site is planned to provide public access to existing and future maps.

2.3. HAWAII

Hazard Assessment funding was first received by Hawaii in 1999, 2 years after the NTHMP was first funded in 1997. Hawaii has the longest state record of tsunami hazard assessment and mitigation efforts, due to a long history of deaths and property loss inflicted by destructive tsunamis. Prior to establishment of the NTHMP, inundation modeling was performed using relatively simple 1-dimensional models and, for decades, evacuation maps based on these model results have been published for each area in public telephone books. Because of this, the Hawaii State Civil Defense Division has been allocating most of the funding to far-field tsunami forecasting efforts, the identification and study of potential local tsunami sources, and instrumentation for early detection and warning of local tsunamis. Some additional support was provided for two inundation modeling efforts, West Honolulu and the Kona Coast of Hawaii, but these unpublished studies are considered preliminary in nature, and have not been used to develop evacuation maps.

Later studies with 2-dimensional models (Wei *et al.*, 2003) demonstrated that the existing evacuation maps at Kaena-Haleiwa and Haena-Hanalei, based on 1-dimensional model simulations, underestimate the 100-year inundation limits. Because of this, a systematic program to update the existing tsunami inundation maps was initiated in 2003. In mid-2003, the Hawaii State Department of Land and Natural Resources agreed to assume responsibility for geotechnical guidance and evacuation mapping in support of the effort as part of the Hawaii Hazard Assessment team, and an official of that agency is now a member of the NTHMP Steering Group.

The Hawaii approach to hazard assessment focuses on the simulation of past events. Consequently, the availability of runup estimates for verification of model results is an important factor in assigning priority to a specific area. Initial sites for inundation modeling and updates of evacuation maps include Kaiaka Bay, Hawaii Kai, and Kaena-Haleiwa, on Oahu; Kahului Harbor, Maui; Haena-Hanalei, Kauai; and Kailua Bay and Hilo Bay, Hawaii. These first priorities were set primarily on the basis of feasibility, i.e., the availability of runup estimates and adequate bathymetric and topographic data. A selec-

Table II. The Alaska Department of Emergency Services organized a review by the Alaska Hazard Assessment team of candidate communities for inundation numerical modeling. The communities were discussed and rated, guided by the criteria indicated.

Community	Tsunami Potential	Community Involvement	Bathymetry[a]	Population	Infrastructure	Tourism	Cruise Ships (Tour Bus/Ship)	Special Seasonal Events	Commercial Fishing/Timber	Large Scale USGS Base Maps
Cold Bay			2	103	✔			✔	✔	
Cordova			3	2571	✔	✔	✔	✔	✔	✔
Homer	✔	✔	1	4155	✔	✔	✔	✔	✔	✔
Seldovia	✔	✔	1	281	✔	✔	✔	✔	✔	✔
Ketchikan		✔	2	8460	✔	✔	✔	✔	✔	✔
King Cove			2	1947	✔			✔	✔	
Petersburg			3	3398	✔	✔		✔	✔	
Sand Point	✔	✔	2	830	✔			✔	✔	
Seward	✔	✔	3	3090	✔	✔	✔	✔	✔	✔
Sitka	✔	✔	2	8779	✔	✔	✔	✔	✔	✔
Unalaska	✔	✔	1	4285	✔	✔	✔	✔	✔	
Valdez	✔		2	4155	✔	✔	✔	✔	✔	✔
Whittier	✔	✔	1	306	✔	✔	✔	✔	✔	✔
Wrangell			2	2589		✔	✔		✔	
Yakutat			1	810	✔	✔		✔	✔	✔
Elfin Cove			2	50		✔			✔	
Ouzinkie	✔		2	252	✔	✔		✔	✔	
Port Lions			2	242	✔	✔		✔	✔	
Akutan			1	408				✔	✔	
Perryville			2	107				✔	✔	
Adak			1–2	7	✔				✔	

Shemya	1	0						
Nikolski		35						
Juneau/Douglas	3	30684	✓	✓	✓	✓	✓	✓
Skagway	3	814	✓	✓	✓	✓		✓
Craig	3	2145	✓	✓		✓	✓	✓
Haines	3	1463	✓	✓	✓	✓	✓	✓

[a] Bathymetry data availablity scale: 1 – good, 2 – some, 3 – poor.

tion and prioritization procedure is currently under development that would include other important factors, such as population and infrastructure at risk.

A website is planned to provide public access to completed maps and reports.

2.4. OREGON

This state has a pre-NTHMP history of active efforts to model the inundation of coastal communities and produce evacuation maps and other emergency products (Priest, 1995). Oregon first received NTHMP Hazard Assessment funding in 1997. The NTHMP partner agencies are Oregon Emergency Management (OEM) and Oregon Department of Geology and Minerals Industry (DOGAMI). Inundation modeling for Oregon has been performed for all but one inundation map by scientists at the Oregon Graduate Institute's School of Science & Engineering at Oregon Health & Science University (OGI) using a Cascadia earthquake source developed in cooperation with DOGAMI (Priest *et al.*, 1997, 2001). The numerical simulation for the City of Gold Beach was completed by the TIME Center. The tsunami inundation maps are produced from the raw numerical simulations by DOGAMI scientists and released as DOGAMI publications. Between 1995 and 1997 three inundation maps were produced. The two inundation maps published in 1997 benefited from NTHMP and TIME Center support that helped to finalize the "credible worst case" tsunami scenario (Priest *et al.*, 1997, 2001). "Credible worst case" source scenarios are typically a large earthquake on the offshore Cascadia Subduction Zone, with an additional concentration of energy in the form of a local asperity. Such scenarios are now the standard for tsunami hazard mapping of open coastal areas of Oregon and Washington. All Oregon inundation maps also depict flooding from a least-severe and moderately-severe local tsunami in order to directly illustrate the uncertainty in the modeling. After 1997, NTHMP funds wholly supported production of five inundation maps. Over the last 5 years DOGAMI and OEM have worked with local government to produce 17 evacuation maps in the form of free brochures. Depiction of evacuation zones is guided by the "credible worst-case" inundation event, although some jurisdictions choose to evacuate to somewhat higher and more inland areas. If an inundation model is not available, DOGAMI scientists estimate the likely inundation based on simulations from nearby areas. Evacuation map brochures and inundation maps are produced in a priority sequence based primarily on population and infrastructure at risk. Maps and related material can be viewed, downloaded, and purchased through the Publications and Data Center link at the DOGAMI web site: http://www.oregongeology.com/.

Information on tsunami simulations, including some animations, can be accessed at the OGI web site: http://www.ccalmr.ogi.edu/projects/oregonian/.

2.5. WASHINGTON

The Emergency Management Division directs Hazard Assessment in Washington State, and the Department of Natural Resources (DNR) provides geotechnical guidance, evacuation map production, and distribution of maps and related reports through the Division of Geology and Earth Resources. The effort was initiated in 1997 when an agreement was reached with the Oregon Graduate Institute for a northward extension of the area being modeled for Oregon, to include the southwest Washington coast. In this manner, inundation maps for four counties were produced – Pacific, Grays Harbor, Clallam, and Jefferson – and evacuation maps were completed for the first two of these counties.

Priorities are now set by consideration of a number of factors, but especially the at-risk population and infrastructure, tsunami potential, availability of bathymetric and topographic data, and community acceptance and involvement. Since 2001, Washington State inundation modeling has been provided by the NOAA TIME Center. An additional inundation map has been completed for the Seattle area, and modeling efforts are underway for the eastern Straits of Juan de Fuca and the Tacoma area. Evacuation map production is in progress for Jefferson County and Clallam County (communities of Neah Bay, La Push, Clallam Bay, Port Angeles, and Sequim) and revisions are planned for Pacific County (Ilwaco/Long Beach, South Bend/Raymond, Grayland/North Cove/Tokeland, Ocean Park/Bay Center) and Grays Harbor County (West Port, Ocean Shores, Hoquiam, Aberdeen/Cosmopolis, Ocean City/Pacific Beach/Taholah). "Credible worst case" source scenarios for areas on the Pacific coast and the Straits of Juan de Fuca are similar to those for Oregon, i.e., a Cascadia Subduction Zone earthquake with a local asperity. For areas in Puget Sound, "credible worst case" sources are developed as earthquakes on the Seattle Fault and other fault systems and, though none have yet been modeled, landslides and delta failures are also under consideration. Existing maps and related publications can be viewed and downloaded at the website of the DNR Division of Geology and Earth Resources: http://www.dnr.wa.gov/geology/.

3. Impact of the Hazard Assessment Program

The impact of a tsunami inundation map on emergency management (EM) officials and citizens alike cannot be overestimated – it is a clarifying, galvanizing catalyst for action. The Hazard Assessment Program in general, and the tsunami inundation maps in particular, have had a major positive impact in the following areas.

Improved collaboration of R&D and EM communities. Because the academic scientists and emergency managers are well-respected and influential members of their respective communities, their vigorous collaboration on hazard mitigation issues has had an important positive impact on the relationship of the tsunami R&D and EM communities. This is a direct result of the successful NTHMP effort to improve this country's tsunami modeling and mapping infrastructure.

Improved planning. Once a map is completed and available for study, previously vague concerns and abstract issues are suddenly and immediately clarified and rendered concrete. It is at this moment that effective, community-specific planning has truly begun – individual hazards can be identified and mitigation measures can be developed and implemented that are specific to that hazard. A map is thus the fundamental starting point for any effective planning and mitigation program, aiding the evaluation of critical issues such as population and infrastructure vulnerability, and the identification of feasible evacuation routes.

Improved education and preparedness. The maps are an absolutely essential educational tool and, to judge the bottom-line impact of these maps, one has to consider their effect on the final users – citizens residing in small and large coastal communities at risk to tsunamis. Once completed, public workshops and informational forums are held to present the maps to citizens of these communities, and to provide an opportunity for discussion of the result with Local, State, and Federal Emergency Managers and the scientists that developed the maps. Again and again, the powerful impact of these maps is made clear – the awareness of a citizen, previously vague and uncertain, dramatically intensifies and, in many cases, prompts the individual to become an active participant in the mitigation program.

Improved survival. Lives will undoubtedly be saved because of the dramatic impact these maps have made on communities. Improvements in emergency planning and preparation, and a more aware and educated population will translate into many fewer fatalities when the next destructive tsunami attacks a U.S. coastal community.

4. Evolving Strategies for Hazard Assessment

Experience and lessons learned during the course of Hazard Assessment work have resulted in the evolution and clarification of some basic principles involving organizational structure, scientific strategy and philosophy, and recommended practices and procedures.

4.1. ORGANIZATIONAL STRUCTURE

In all states except Oregon, a state Emergency Management official directs the Hazard Assessment effort, selects and provides financial support to a numerical modeling scientist, and serves on the decision-making, eight-member NTHMP Executive Committee. In Oregon, contracting for the modeling and hazard assessment is managed by the state Geotechnical Agency, which shares voting privileges on the Executive Committee with the state Emergency Management official. In all states, the role of a state Geotechnical Agency official includes working closely with Emergency Management and the numerical modeling scientist to address the technical feasibility, potential value, and scientific credibility of a proposed effort; to set priorities; to interpret the modeling results; and to develop, publish, and distribute evacuation maps and other emergency management products. Priorities are established by state officials through a systematic evaluation of relevant factors pertaining to each coastal community, including the history of tsunami events, scientific assessments of future tsunami potential, population and infrastructure characteristics and needs, an interest and willingness to participate and contribute to the effort, and technical feasibility.

As part of the organizational structure, the NOAA TIME Center provides scientific and technical guidance and assistance to all states, including close collaboration with state-contracted modelers; importing and implementing technological advances, tools, and methods to improve scientific quality and production (e.g., Mofjeld *et al.*, 2003); and the development of standards and quality control procedures. Finally, the organizational structure also includes a Technical Advisory Committee (TAC), which is tasked with advising the NTHMP Steering Group on technical issues related to tsunami research, development, operations, and mitigation.

4.2. SCIENTIFIC STRATEGY AND PHILOSOPHY

The NTHMP emphasizes development, implementation, and application over research, with the focus being on the timely delivery of scientifically credible and useful products. With respect to Hazard Assessment, it is now widely acknowledged that numerical modeling technology is sufficiently advanced to produce inundation maps that are valuable for emergency management (Bernard and González, 1994). However, perfect numerical models do not exist. Moreover, potential improvements to a specific numerical model application can almost always be identified – frequently, for example, development of a better quality bathymetric/topographic grid is considered as a possible improvement.

In practice, the additional time needed to develop, implement, test, and evaluate the effects of a modification that might improve model results can force Hazard Assessment teams to make difficult judgments and decisions that involve trade-offs between quality and production. Ever higher quality and utility are natural goals for a scientist, but these must be tempered by the reality of limited time and resources. Rapid and complete coverage of all coastal communities is a natural goal for an emergency manager, but this must be tempered by the need for scientifically credible and defensible products. Ultimately, the final decision must be made by the State Emergency Management official on the basis of recommendations by the scientific and technical members of the team. There is no universal recipe for making such decisions because every Hazard Assessment effort is different but, to aid this process, the Hazard Assessment component has developed a set of "Recommended Practices and Procedures."

4.3. RECOMMENDED PRACTICES AND PROCEDURES

Some useful guidelines have emerged in the course of work conducted by the NOAA TIME Center to perform Washington inundation modeling and to assist Alaska in an effort to accelerate inundation mapping. Based on this experience, "Recommended Practices and Procedures" to guide modeling efforts have been developed, and are presented in Table III.

This guidance is needed because each Hazard Assessment effort is a complex, labor-intensive enterprise, requiring the collaborative effort of a sizable Hazard Assessment team: state Emergency Management officials, Geotechnical Agency scientists, numerical modelers, and NOAA TIME Center personnel. State and other geoscientists must help define appropriate sources. An effective technical infrastructure is essential, and must include bathymetric and topographic data, personnel skilled in the development of acceptably accurate bathymetric/topographic digital elevation models, adequate computational capabilities, and hardware and software tools for display, quality control, and analysis. Professional judgments must also be made before model development as to the probable success of the effort – i.e., whether the result will be both scientifically credible and useful. This includes an assessment of factors such as the quality and availability of bathymetric and topographic data, knowledge of potential sources, and the physical complexity of the region. Finally, as the work progresses, judging the trade-offs between quality and production may be required, and a decision must be made on the appropriate time to stop seeking and implementing potential improvements, to complete and document the work, and to begin the next effort.

5. Improvements

After the formal review of the NTHMP in 2001, reviewers submitted comments that suggested a number of improvements to the Hazard Assessment component. The following are the areas suggested for improvement and ongoing NTHMP efforts to address these issues.

Standards. Reviewers were concerned that each institution uses different numerical models that may give different results, and suggested that quality control indices and recommended practices and procedures should be developed to ensure that minimum standards are met. This paper has outlined a preliminary set of recommended practices and procedures to address this issue (Table III).

Improve bathymetric and topographic database. Reviewers approved of NTHMP plans to strengthen relationships with the National Ocean Service (NOS) and USGS. These agencies have been advised of NTHMP data priorities and now consider these as a factor in scheduling data collection and processing. NOS has been especially responsive to data priorities of the NOAA TIME Center with regards to data collection and processing in

Table III. Recommended practices and procedures for inundation modeling efforts conducted by the U.S. National Tsunami Hazard Mitigation Program

1. *A State Emergency Management official* should direct the effort, bearing ultimate responsibility for setting priorities
2. *A State Geotechnical official* (State equivalent of a USGS official) should serve as the State technical adviser to the State Emergency Management official, contribute to technical issues of the modeling effort, and bear responsibility for development and publication of inundation and evacuation maps
3. *Timely delivery of scientifically credible products* should drive the effort, with development taking priority over research
4. *Community Prioritization* should involve a systematic process of assessing community characteristics to rank need and potential success of a Hazard Assessment effort. The prioritization matrix developed by the Alaska Tsunami Mapping Team is an excellent example (Table II)
5. *Source Identification* should involve an intensive, highly focused workshop to systematically inventory and document the best available scientific information on potential tsunami sources in a region. An example of this is the recent Puget Sound Tsunami Sources Workshop jointly organized by Washington State, the USGS, and NOAA (González *et al*., 2003)
6. *Computational Grid Development* should (a) aim to create fine-scale inundation grids of 50 m resolution or less if important, site-specific inundation features are to be identified (Summary Report of 6–8 November 2001 Meeting of the NTHMP Steering Group, see http://www.pmel.noaa.gov/tsunami-hazard/Summaries.html), (b) conduct an intensive effort to acquire State and local data, and (c) resolve vertical and horizontal datum differences in a careful, systematic manner
7. *Milestone Meetings* of the entire Hazard Assessment team should make and document team decisions that are needed to achieve key milestones. To ensure that milestones are actually achieved at these formal meetings, frequent informal communication between collaborating team members is essential. Some recommended milestones, advance materials required, and issues that call for team decision-making are as follows

Table III. (Continued)

– *Milestone*: Define scope of the effort

- Material: Inventory of bathymetric and topographic data
  Potential tsunami source information
- Issues: Adequacy of available bathymetric and topographic data
  Geographical extent of computational grid system
  Communities to be modeled at the finest resolution
  Source specification
  Judgment of scientific credibility and utility of proposed effort
  Deliverable products
  Schedule

– *Milestone*: Critique preliminary model runs

- Material: Graphics of results
  Modeler's brief interpretive narrative and recommendation
- Issues: Assessment of scientific credibility and utility
  General reasonableness
  Consistency with available historic and pre-historic data
  Impact of proposed improvements on resources and schedule
  Production plan and schedule

– *Milestone*: Critique products

- Material: Deliverable products in graphic and tabular form
- Issues: Impact of modifying or adding products
  Final acceptance
  Documentation plan
  Schedule for delivery

8. *Inundation modeling products* should be provided as hard-copy graphics and digital data and graphic files on a CD-ROM that are GIS-compatible, and should include (e.g., Titov *et al.*, 2003; also see Figure 1):
   Animation of Simulations
   Shoreline Vectors
   Bathymetric/Topographic Grids
   Source Deformation
   Maximum Wave Heights
   Maximum Water Depth on Land
   Maximum Current Speeds
   Inundation Line Vectors
   Zoned Maximum Water Depth on Land (Low, Med, Hi)
   Zoned Maximum Current Speeds (Low, Med, Hi)
   Time series at selected stations
   ArcView Project Module

Table III. (Continued)

Metadata files

Documentation in the form of a Product Report

9. *A Final Inundation Modeling Study Report* should be provided to the State after the acceptance and delivery of all digital and graphical products (e.g., Titov et al., 2003; also see Figure 1).
10. *Inundation and evacuation maps* based on the inundation modeling study should be developed and published by the State (e.g., Walsh *et al.*, 2003b; also see Figures 2 and 3), and consultations with the inundation modeler should be available to the State during this effort, as needed, to clarify technical issues.
11. *Archival copies* of all modeling and mapping products should be delivered to the NOAA TIME Center.

Alaskan waters. The TIME Center is also exploring special processing methods by the USGS to provide topographic products.

Landslide sources. Recent historical events strongly suggest that subaerial and subaqueous landslide events can be a significant source of destructive tsunamis, and modeling such events is a high priority. Alaska and, to some degree, Hawaii have a history of destructive tsunamis generated by landslides, and the University of Alaska at Fairbanks plans to include landslides in future modeling efforts. In Washington, several such potential sources were identified during the Puget Sound Tsunami Sources Workshop (González *et al.*, 2003), and the Washington Hazard Assessment team will likely use a landslide source for planned inundation model simulations of the Tacoma area. Finally, a high priority is given to modeling the impact of landslide tsunami sources that may exist on the continental slope of northern California, Oregon, and Washington, an area at the epicenter of Cascadia Subduction Zone earthquakes of magnitude 9.

Impact modeling. Estimates of the destructive impact of a tsunami on structures and humans is valuable information. The NTHMP Mitigation Subcommittee organized and conducted a workshop to address this issue (Walsh *et al.*, 2003a). The new NSF Network for Earthquake Engineering Simulation (NEES) Program has funded the construction of a new tsunami wave tank facility at Oregon State University (http://www.eng.nsf.gov/nees/); the facility became available for research in September 2003 (http://wave.oregonstate.edu/). Also, the NEES grand challenge report for the National Research Council establishes a coordinated effort between NTHMP and NEES. The highest priority item is to use the new NEES facility to determine forces on structures for construction guidance.

Probabilistic methods. Current inundation maps are based on the development of a "credible worst case" scenario. (The exception is Hawaii, as indicated in Section 2.3, above.) While maps based on this concept are useful, emergency managers are now expressing a need for probabilistic estimates. Conceptually, probabilistic inundation maps could be produced through multiple tsunami model simulations based on probabilistic source information. This approach may be possible for far-field sources, but would be difficult for local sources, especially landslides. But performing multiple simulations of an area could produce information on the sensitivity of that area to a variety of credible sources. Although probabilities could not be assigned, this information would add value to the results of a single, credible worst case simulation.

6. Summary

This report provides a summary of progress to date by the NTHMP Hazard Assessment component. Accomplishments include the production and publication of 22 inundation maps that cover 113 coastal communities with an estimated combined population of more than 1.2 million residents that are at risk.

Achieving the first Hazard Assessment goal of providing every at-risk community with an inundation map, an evacuation map, and related Hazard Assessment products, remains a major challenge. Practices and procedures must continue to be improved in order to increase the current rate of production while maintaining scientific credibility. The recent addition to the Steering Group of Geotechnical Agency representatives for two States is an improvement in the organizational structure that should also help increase map production.

Scientific and technical improvements are also needed, including improvements in the bathymetric and topographic database, the development and implementation of probabilistic methods, and estimates of tsunami impact on structures and humans. The NTHMP is coordinating research and development priorities with the wider community, especially the NSF NEES Program. Future research results will be implemented and, as these improvements are realized, a formal program must be established for the systematic review and improvement of existing inundation and evacuation maps.

References

Bernard, E. N.: 2001, The U.S. National Tsunami Hazard Mitigation Program Summary. In: *Proceedings of the International Tsunami Symposium 2001* (*ITS 2001*) (on CD-ROM), NTHMP Review Session, R-1, Seattle, WA, 7–10 August 2001, pp. 21–27. http://www.pmel.noaa.gov/its2001/.

Bernard, E. N. and González, F. I.: 1994, Tsunami Inundation Modeling Workshop Report (November 16–18, 1993). Technical Report NOAA Tech. Memo. ERL PMEL-100 (PG94-143377), NOAA/Pacific Marine Environmental Laboratory, Seattle, WA.

Crawford, G. L.: 2001, Tsunami inundation preparedness in coastal communities. In: *Proceedings of the International Tsunami Symposium 2001* (*ITS 2001*) (on CD-ROM), NTHMP Review Session R-18, Seattle, WA, 7–10 August 2001, pp. 213–219. http://www.pmel.noaa.gov/its2001/.

Eisner, R. K., Borrero, J. C., and Synolakis, C. E.: 2001, Inundation maps for the State of California. In: *Proceedings of the International Tsunami Symposium 2001 (ITS 2001)* (on CD-ROM), NTHMP Review Session, R-4, Seattle, WA, 7–10 August 2001, pp. 67–81. http://www.pmel.noaa.gov/its2001/.

Fryer, G. J., Cheung, K. F., Smith, Jr., J. R., Teng, M. H., and Watts, P.: 2001, Inundation mapping in Hawaii. In: *Proceedings of the International Tsunami Symposium 2001 (ITS 2001)* (on CD-ROM), NTHMP Review Session, R-16, Seattle, WA, 7–10 August 2001, p. 207. http://www.pmel.noaa.gov/its2001/.

González, F. I., Sherrod, B. L., Atwater, B. F., Frankel, A. P., Palmer, S. P., Holmes, M. L., Karlin, B. E., Jaffe, B. E., Titov, V. V., Mofjeld, H. O., and Venturato, A. J.: 2003, Puget Sound Tsunami Sources – 2002 Workshop Report, A Contribution to the Inundation Mapping Project of the U.S. National Tsunami Hazard Mitigation Program. NOAA OAR Special Report, NOAA/OAR/PMEL, 34 pp.

González, F. I., Titov, V. V., Mofjeld, H. O., Venturato, A. J., and Newman, J. C.: 2001, The NTHMP Inundation Mapping Program. In: *Proceedings of the International Tsunami Symposium 2001 (ITS 2001)* (on CD-ROM), NTHMP Review Session, R-2, Seattle, WA, 7–10 August 2001, pp. 29–54. http://www.pmel.noaa.gov/its2001/.

Hansen, R., Suleimani, E., Kowalik, Z., and Combellick, R.: 2001, Tsunami inundation mapping for Alaska Communities. In: *Proceedings of the International Tsunami Symposium 2001 (ITS 2001)* (on CD-ROM), NTHMP Review Session, R-12, Seattle, WA, 7–10 August 2001, p. 181. http://www.pmel.noaa.gov/its2001/.

Mofjeld, H. O., Venturato, A. J., González, F. I., Titov, V. V., and Newman, J. C.: 2003, The harmonic constant datum method: Options for overcoming datum discontinuities at mixed/diurnal tidal transitions. *J. Atmos. Oceanic Tech.* **21**, 95–104.

Priest, G. R.: 1995, Explanation of mapping methods and use of the tsunami hazard maps of the Oregon coast. Oregon Department of Geology and Mineral Industries Open-File Report O-95-67, 95 pp.

Priest, G. R., Baptista, A. M., Fleuck, P., Wang, K., Kamphaus, R. A., and Peterson, C. D.: 1997, Cascadia subduction zone tsunamis: Hazard mapping at Yaquina Bay, Oregon. Final technical report to the National Earthquake Hazard Reduction Program: Oregon Department of Geology and Mineral Industries Open-File Report O-97-34, 144 pp.

Priest, R., Baptista, A. M., Meyers, E. P., III, and Kamphaus, R. A.: 2001, Tsunami hazard assessment in Oregon. In: *Proceedings of the International Tsunami Symposium 2001 (ITS 2001)* (on CD-ROM), NTHMP Review Session, R-3, Seattle, WA, 7–10 August 2001, pp. 55–65. http://www.pmel.noaa.gov/its2001/.

Suleimani, E. N., Combellick, R. A., Hansen, R. A., and Carver, G. A.: 2002a, Tsunami hazard mapping of Alaska coastal communities. State of Alaska Division of Geological and Geophysical Surveys, Alaska Geosurvey News, Vol. 6, No. 2, June 2002 (download at http://www.dggs.dnr.state.ak.us).

Suleimani, E. N., Hansen, R. A., Combellick, R. A., Carver, G. A., Kamphaus, R. A., Newman, J. C., and Venturato, A. J.: 2002b, Tsunami hazard maps of the Kodiak area, Alaska. State of Alaska Department of Natural Resources – Division of Geological and Geophysical Surveys, Alaska, RI 2002-1, 18 pp.

Titov, V. V., González, F. I., Mofjeld, H. O., and Venturato, A. J.: 2003, NOAA TIME Seattle Tsunami Mapping Project: Procedures, data sources, and products. Technical Report NOAA Tech. Memo. OAR PMEL-124, Seattle, WA.

Tsunami Hazard Mitigation Federal/State Working Group: 1996, Tsunami Hazard Mitigation Implementation Plan – A Report to the Senate Appropriations Committee. PDF download at http://www.pmel.noaa.gov/tsunami-hazard/, 22 pp., Appendices.

Walsh, T. J., Crawford, G. L., Eisner, R., and Preuss, J. V.: 2003a, Proceedings of a Workshop on Construction Guidance for Areas of High Seismic and Tsunami Loading. Washington Military Department, 25 pp.

Walsh, T. J., Titov, V. V., Venturato, A. J., Mofjeld, H. O., and González, F. I.: 2003b, Tsunami hazard map of the Elliott Bay area, Seattle, Washington—Modeled tsunami inundation from a Seattle fault earthquake. 1 plate, scale 1:50,000.

Wei, Y., Cheung, K. F., Curtis, G. D., and McCreery, C. S.: 2003, Inverse algorithm for tsunami forecasts. *J. Waterw. Port Coast. Ocean Eng.* **129**(3), 60–69.

Natural Hazards (2005) 35: 111–119

Local Tsunami Warning in the Pacific Coastal United States

MARK DARIENZO[1]*, AL AYA[2], GEORGE L. CRAWFORD[3], DAVID GIBBS[4], PAUL M. WHITMORE[5], TYREE WILDE[6] and BRIAN S. YANAGI[7]

[1]*Department of Land Conservation and Development, 635 Capitol St., Suite 150, Salem, OR 97301-2540, USA;* [2]*Cannon Beach Rural Fire Protection District, P.O. Box 24, Cannon Beach, OR 97110, USA;* [3]*Washington State Military Department, Emergency Management Division, M/S TA-20, Camp Murray, WA 98430-5211, USA;* [4]*Kenai Peninsula Borough, Office of Emergency Management, 144 N. Binkley St., Soldotna, AK 99669, USA;* [5]*West Coast Alaska Tsunami Warning Center, 910 S. Felton St., Palmer, AK 99645, USA;* [6]*National Weather Service, 5241 NE 122nd Ave., Portland, OR 97230-1089, USA;* [7]*Hawaii Civil Defense, 3949 Diamond Head Rd., Honolulu, HI 96816-4495, USA*

(Received: 28 August 2003; accepted: 22 March 2004)

Abstract. Coastal areas are warned of a tsunami by natural phenomena and man-made warning systems. Earthquake shaking and/or unusual water conditions, such as rapid changes in water level, are natural phenomena that warn coastal areas of a local tsunami that will arrive in minutes. Unusual water conditions are the natural warning for a distant tsunami. Man-made warning systems include sirens, telephones, weather radios, and the Emergency Alert System. Man-made warning systems are normally used for distant tsunamis, but can be used to reinforce the natural phenomena if the systems can survive earthquake shaking. The tsunami warning bulletins provided by the West Coast/Alaska and Pacific Tsunami Warning Centers and the flow of tsunami warning from warning centers to the locals are critical steps in the warning process. Public knowledge of natural phenomena coupled with robust, redundant, and widespread man-made warning systems will ensure that all residents and tourists in the inundation zone are warned in an effective and timely manner.

Key words: tsunami warning, warning system, siren, telephone, weather radio, emergency alert system, natural tsunami warning, ground shaking, slow earthquake, submarine landslide, evacuation, population at risk, sea level, tsunami watch, tsunami warning

Abbreviations: AOR – area of responsibility, ANI – Automatic Number Information, EAS – Emergency Alert System, EMWIN – Emergency Managers Weather Information Network, FEMA – Federal Emergency Management Agency, NADIN2 – National Airspace Data Interchange Network, NAWAS – National Warning System, NOAA – National Oceanic and Atmospheric Administration, NWR – NOAA Weather Radio, NWS – National Weather Service, NWWS – National Weather Wire Service, PSAPS – Public Service Answering Points, PTWC – Pacific Tsunami Warning Center, SAME – Specific Area Message Encoding, SAWAS – State Warning System, WC/ATWC – West Coast/Alaska Tsunami Warning Center

* Author for correspondence: Tel: +1-503-373-0050 ext. 269; Fax: +1-503-378-5518; E-mail: mark.darienzo@state.or.us

1. Introduction

Tsunamis can be destructive to coastal areas. Warning about impending arrival can save thousands of lives (Figure 1). From the point of view of warning and evacuation, the origin of a tsunami is either local or distant. Local tsunamis arrive at the coast in minutes, while distant tsunamis arrive at the coast in hours. For example, a tsunami generated by an Alaska subduction zone earthquake would be local for Alaska but distant to California, Oregon, Washington, and Hawaii. Coastal communities are warned of a tsunami either through natural phenomena that they can feel or witness or by warning systems, such as sirens or weather radios, triggered once messages are received from Tsunami Warning Centers.

Man-made warning systems, such as sirens, telephones, weather radios, and the Emergency Alert System (EAS), are most effective for warning of tsunamis from distant sources. Man-made systems are also useful for the "all clear" notification for distant tsunamis and, if the equipment survives the earthquake, for local tsunamis. Strong earthquake shaking and unusual water conditions, such as rapid drawdown or sudden rise of the ocean, are natural warnings for a tsunami from local sources. Evacuation must be immediate. A community's man-made warning system could be damaged by the earthquake and, therefore, unavailable as a warning tool. If communities

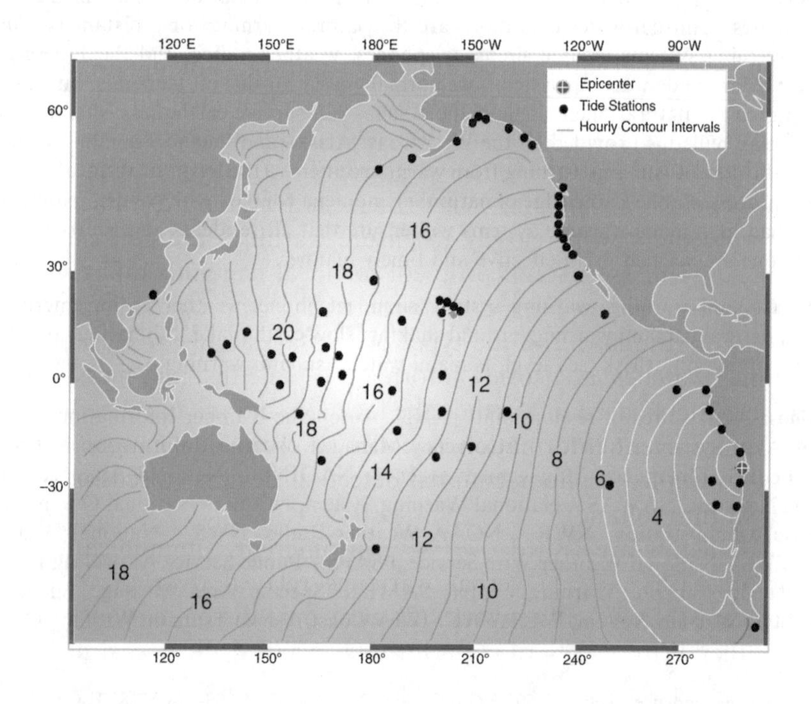

Figure 1. Tsunami arrival time map (each contour = 1 hour)

desire a man-made system to reinforce the natural phenomenon, the system must be hardened and provided a reliable power source.

The paper focuses on how the public is warned of an impending tsunami, how effective are the means of warnings, and how they can be improved.

2. Tsunami Warning for Distant Tsunamis

2.1. TSUNAMI WARNING CENTERS

The Tsunami Warning Centers, operated by the National Oceanic and Atmospheric Administration's (NOAA) National Weather Service (NWS), have primary responsibility to issue tsunami watches, warnings, and other bulletins to governments and populations within their areas of responsibility (AOR), especially to emergency officials. Local emergency officials are responsible for coordinating and managing local warning and evacuations. In other words, the warning centers warn emergency officials. Local emergency officials notify and advise local populations to evacuate.

The two Tsunami Warning Centers for the United States and Canada are the West Coast/Alaska Tsunami Warning Center (WC/ATWC) and the Pacific Tsunami Warning Center (PTWC). The WC/ATWC, located in Palmer, Alaska, issues tsunami warnings to coastal residents of California, Oregon, Washington, British Columbia, and Alaska. The PTWC, located in Ewa Beach, Hawaii, is responsible for Hawaii, U.S interests throughout the Pacific basin, and international coordination during tsunami events. The PTWC and WC/ATWC monitor seismic events throughout the Pacific Ocean. They also interact with regional and national centers monitoring seismological and tidal stations for the purpose of evaluating an earthquake's potential to generate a tsunami. When there is a regional or Pacific-wide event, the two centers work closely to ensure warnings are consistent. Although the text of messages and protocol for local tsunamis differ slightly, the general criteria each center uses to determine the tsunami-producing potential of an earthquake and the types of messages sent are similar. Tsunami warnings, watches, advisories, information bulletins, and messages are issued by the centers based on earthquake location and magnitude. They are updated as more event information is collected from tide gauges and the more recently installed seafloor tsunami detection devices (discussed in another paper in this volume).

Tsunami bulletins are issued over several different communication systems. Primary paths are (1) verbal warnings relayed over the National Warning System (NAWAS) network, (2) electronic dissemination over the NOAA Weather Wire, (3) hard-copy through NWS dedicated circuits and the Federal Aviation Administration's National Airspace Data Interchange Network (NADIN2) communication system, and (4) the NWS Emergency

Managers Weather Information Network (EMWIN). NWS offices also disseminate these messages over NOAA Weather Radio (NWR). The EAS is activated through the NWS forecast offices and/or the state departments of emergency services. Individual state EAS plans determine which tsunami messages and associated message type headers will activate local EAS decoders. State emergency service offices, Federal Emergency Management Agency (FEMA), military contacts, U.S. Coast Guard, and NWS offices are primary sites to receive tsunami bulletins. NAWAS messages are sent to coastal Public Service Answering Points (PSAPs) via SAWAS (State Warning System) by the state emergency management office. Secondary methods of message dissemination are direct phone contacts, e-mail messages, home page updates (http://wcatwc.gov), and experimental pager notification system.

For a distant tsunami, the decision to evacuate all low-lying areas rests on local jurisdictions. They activate their local community warning system through pre-planned procedures once they receive the warning from the Tsunami Warning Centers.

2.2. MANMADE WARNING SYSTEMS

Various types of systems for notifying coastal residents and visitors of tsunami warnings issued from the warning centers are available and discussed below. Systems include sirens, telephones, NOAA weather radios, and the EAS. The goal of any community is to have the most effective coverage for the lowest cost. Costs of implementing new systems can be relatively high. Costs include the major elements, as well as labor, maintenance, and training. However, new systems are more reliable than older ones. High costs of new systems may be offset by higher maintenance costs and relatively low efficiency of older systems. If several adjacent communities use the same system, cost savings result from equipment purchased in quantity (Oregon Emergency Management and Oregon Department of Geology and Mineral Industries, 2001).

2.2.1 *Sirens*

Sirens are either electro-mechanical or fully electronic devices (Federal Emergency Management Agency, 1980) (Figure 2). They can be triggered manually or automatically. Both types project standard siren sounds, but electronic sirens can also broadcast public address announcements. Public address announcements should be concise and, ideally, pre-recorded to avoid potential problems with unintelligible messages from a stressed system operator.

Sirens should have only one tone that sounds for a distinct period of time. This standardization reduces confusion for resident or transient populations along the coast. The siren normally would prompt people to turn on their radio or television for more information. Relative to other types of man-made warning systems, sirens have the advantage of reaching all populations in the

Figure 2. Siren in Hawaii (Hawaii Civil Defense).

coverage area. Sirens are most useful in areas such as crowded beaches, where access to warning devices, such as radio or television, is limited. A major barrier to installation of sirens for many communities is cost. Ultimately, the siren system adopted by a community depends on the amount of funds available and the type and area of coverage a community needs and wants.

2.2.2 *Telephones*

Telephone warning systems use Enhanced 9-1-1 ANI (Automatic Number Information) data to call telephone subscribers within a designated area (homes and businesses) and give call recipients a pre-recorded telephone message with evacuation or other relevant event information. Telephone alerts can be activated by the emergency operation center (either local or central) or local public safety dispatch centers.

Telephones can reach large audiences quickly with a prerecorded message. At present cellular phones are not connected to the system. It can be used for multiple hazards. There are limitations. You must be near a telephone. Phone lines might not be functional if damaged by an earthquake. The lines could become overloaded if people call to confirm what they heard about a distant tsunami threat or to confirm that the shaking felt is from a tsunami-producing earthquake. Not all answering machines are designed to pick up the recorded message, although the system is designed to call back two more times.

2.2.3 *Weather Radios*

National Weather Service broadcasts all its warnings, including tsunami warnings, on its existing VHF and UHF network, known as NOAA Weather

Radio. Upon broadcast, tsunami warnings activate alarms on specially designed weather radio receivers. The entire network was recently upgraded to interface with the new EAS discussed in a later section. EAS and NOAA Weather Radio use a digital encoding scheme known as Specific Area Message Encoding (SAME). SAME allows a warning to be broadcast over a large area while activating alarms on only certain receivers. Users program their receivers for certain types of alarms considered relevant for their specific location.

A warning broadcast on NOAA Weather Radio can be automatically rebroadcast on commercial radio, television, and cable. Local authorities could use the NWR system (directly or indirectly) to provide urgent follow-up information to the local public after the tsunami warning is broadcast. The warning issued via NOAA Weather Radio may include a message such as "take appropriate action or follow instructions from local authorities." This will be effective if the public is properly educated. However, state and local emergency management can send out other information over NOAA radio, such as a call to evacuate to areas in need of evacuation information. The state can either (1) request that NWS send out an evacuation notice or (2) set up a system (requiring extra equipment) that allows a state or local jurisdiction to send an audio EAS message to NWS. The message would either be automatically sent to the targeted areas or reviewed by NWS and then transmitted to the targeted areas. NOAA Weather Radio receivers are highly portable and a tourist population could be encouraged to carry them when away from home. Operators of hotels, restaurants, conference centers, campgrounds, and others should be encouraged to have a receiver on site where a responsible authority can monitor it (e.g., the front desk, camp host, and so on). However, there are some limitations to using NOAA Weather Radio as a tsunami warning system. Tourist populations would not be served unless the lodging facilities (hotel, motels, campgrounds), restaurants, or stores have weather radio receivers. In addition, tourists visiting isolated beaches or other low-lying areas do not normally have a weather radio. Not all areas are within range of a transmitter, because of the rugged nature of some coastlines. The existing transmitters require a line of sight for adequate reception.

2.2.4 *Emergency Alert System (EAS)*

In 1996, EAS replaced the outdated Emergency Broadcast System. EAS is now the nation's primary method for notifying the general public of emergency for all hazards. EAS stations include all radio, television, and large cable operators. Messages can originate from any designated authority: from the nation's president, to the state governor, down to the local incident commander.

NOAA/NWS receives the WC/ATWC tsunami watch or warning, rewords the message for media purposes, and retransmits it over NOAA

Weather Radio with SAME encoding, which is used to activate EAS as well as NOAA weather radios. The state receives the watch or warning over NAWAS or National Weather Wire Service (NWWS) and transmits the message over EAS. Local emergency managers, with their own equipment, can insert messages into the EAS network. In essence they can simultaneously commandeer all broadcast facilities within a given area. Local broadcasters and emergency managers must work together to develop and exercise a local and state EAS plan. Once the plan is in place the broadcasters must carry messages specified in the plan.

A key advantage to the EAS system is the flexibility it allows local authorities to develop their own emergency messages. EAS can be used to provide multiple announcements, including evacuation notices, once a tsunami warning has been broadcast. Message can be prerecorded and stored for immediate use.

Critical problems with EAS are the existing and potential non-participation by broadcasters and the need for continuous education and testing. At present, the system should not be considered fully reliable. Many local jurisdictions have no functioning plans and are not actively pursuing them.

3. Tsunami Warning for Local Tsunamis

Strong ground shaking from an earthquake should be the trigger for rapid evacuation from low-lying areas on the coast. If people wait for the other natural warning phenomena (rapid draw down or sudden rise of the ocean) it could be too late to reach safe ground. In the event of a distant tsunami, water changes are a timely natural warning. Defining strong shaking and duration of shaking for coastal residents and tourists is a challenge. These descriptors are highly subjective. One possibility is to use shaking as the trigger for evacuation and err on the side of caution. If "all clear" notifications are made rapidly enough, community disruption by false alarms would be reduced. Another possibility is to leave the descriptor up to local government. Areas with more background seismic activity could choose a higher threshold than those with a lower background.

However, using only strong ground shaking as the trigger for evacuation creates problems when four special situations are considered. The special situations are slow earthquakes, smaller subduction zone earthquakes, inland earthquakes, and earthquake-induced submarine landslides.

3.1. SLOW EARTHQUAKES

Slow subduction zone earthquakes are not usually felt but could still produce a devastating tsunami. Rapid water level changes would be natural warnings for slow earthquakes. A good example is the 7.2 magnitude Nicaragua earthquake of 1992. In the case of Nicaragua, the tsunami arrived about 45 minutes after the earthquake. Timely warning center messages are essential for an event like this. Although tsunami warnings would be issued by the warning centers within 15 minutes of the earthquake, detection of slow earthquakes and determination of their tsunami potential is a challenge.

3.2. SMALL SUBDUCTION ZONE EARTHQUAKES

An earthquake along one short segment of a subduction zone would be felt in adjacent areas and produce a tsunami inundating those areas. The tsunami would arrive relatively quickly, anywhere between 45 minutes to 1–2 hours, depending on distance from the ruptured segment. Once again a warning from the centers will go out within 15 minutes of the earthquake origin time, thus arriving in a reasonable time for evacuation to take place. Some communities educate the public to evacuate with any felt shaking and then quickly send the all clear message if no tsunami danger is found.

3.3. INLAND EARTHQUAKES

Coastal ground shaking does not necessarily indicate a tsunami was generated. The earthquake epicenter could be miles inland from the coastline, but still be felt. Examples are the 1993 Scotts Mills earthquake in Oregon, and the 1999 Satsop and 2001 Nisqually earthquake in Washington. Evacuation is unnecessary and repeated false evacuations are costly not only in terms of lost revenue, but in lost credibility. If there is no tsunami potential, evacuation could be prevented or quickly halted if the coast is notified quickly.

3.4. LANDSLIDES

Local offshore or onshore earthquakes produce landslides, although landslides don't always require an earthquake to trigger them. A submarine landslide or subaerial landslide, which flows into a body of water, can cause a tsunami. The earthquake may or may not be felt. If not felt, unusual water conditions are the warning. A localized tsunami would arrive in minutes. A submarine landslide may have been the cause of a 6 m (19.5 feet) high wave that followed a 5.2 magnitude earthquake in southern California in 1930.

4. Discussion and Recommendations

Either natural phenomena or man-made systems will prompt people to move to safe ground. Natural phenomena can be very subjective. Ground shaking and changes in ocean level may be interpreted incorrectly even by an educated person. Unique situations such as slow earthquakes and landslides add another layer of complexity. Tourist and residential populations constantly change day by day and therefore public education is never-ending.

Man-made warning systems can be costly or individually inadequate to cover all hazard areas. Man-made systems must also be robust enough to withstand a damaging earthquake and redundant if one fails. For example, if a community has a public announcement siren system, they should also have EAS in place to effectively reach as many people as possible. Not only would this redundancy effectively warn communities of the distant tsunami threat, they would supplement the warning from natural phenomena. However, if man-made systems are to be relied upon to reinforce natural phenomena, they must be strengthened to withstand the earthquake.

Quickly knowing the location and magnitude of an earthquake can stop unnecessary evacuation or halt evacuation in progress in the event of non-tsunami-producing earthquakes. Therefore, systems that can provide this service must be readily available. Also, accurate and faster warnings from the tsunami warning centers can reduce or eliminate unnecessary local emergency actions.

Communities will be much safer if coverage is wide, information is quickly disseminated, and natural phenomena are reinforced by robust and redundant man-made warning systems. No matter what system a community has or how state-of-the-art it is, if the public is not familiar with the meaning of the issued alarm, the system is ineffective. Therefore, continuous and effective public education is a key tool in creating tsunami-ready communities.

Acknowledgements

Special thanks to my coauthors for their willingness to review and edit the paper. Thanks also to the anonymous reviewers for their suggestions. Without the support of the National Tsunami Hazard Mitigation Program this paper would not have been written.

References

Federal Emergency Management Agency: 1980, Outdoor Warning Systems Guide. FEMA CPG 1–17.

Oregon Emergency Management and Oregon Department of Geology and Mineral Industries: 2001, Tsunami Warning Systems and Procedures: Guidance for Local Officials. Oregon Department of Geology and Mineral Industries Special Paper 35.

Natural Hazards (2005) 35: 121–139

Planning for Tsunami-Resilient Communities

C. JONIENTZ-TRISLER[1★], R. S. SIMMONS[2], B. S. YANAGI[3], G. L. CRAWFORD[4], M. DARIENZO[5], R. K. EISNER[6], E. PETTY[2] and G. R. PRIEST[7]

[1]*Department of Homeland Security, Federal Emergency Management Agency, Region X, Mitigation Division, 130–228th Street SW, Bothell, WA 98021-9796, USA;* [2]*Alaska Division of Emergency Services, P.O. Box 5750, Fort Richardson, AK 99505-5750, USA;* [3]*Hawaii Civil Defense, 3949 Diamond Head Rd., Honolulu, HI 96816-4495, USA;* [4]*Washington State Military Department, Emergency Management Division, M/S TA-20, Camp Murray, WA 98430-5211, USA;* [5]*Oregon Emergency Management, P. O. Box 14370, Salem, OR 97309-5062, USA;* [6]*CISN and Earthquake Programs, Governor's Office of Emergency Services, 724 Mandana Boulevard, Oakland, CA 94610-2421, USA;* [7]*Coastal 18 Section Leader, Oregon Department of Geology and Mineral Industries, Coastal Field Office, 19 313 SW 2nd, Suite D, Newport, OR 97365, USA*

(Received: 8 September 2003; accepted: 27 April 2004)

Abstract. The National Tsunami Hazard Mitigation Program (NTHMP) Steering Committee consists of representatives from the National Oceanic and Atmospheric Administration (NOAA), the Federal Emergency Management Agency (FEMA), the U.S. Geological Survey (USGS), and the states of Alaska, California, Hawaii, Oregon, and Washington. The program addresses three major components: hazard assessment, warning guidance, and mitigation. The first two components, hazard assessment and warning guidance, are led by physical scientists who, using research and modeling methods, develop products that allow communities to identify their tsunami hazard areas and receive more accurate and timely warning information. The third component, mitigation, is led by the emergency managers who use their experience and networks to translate science and technology into user-friendly planning and education products. Mitigation activities focus on assisting federal, state, and local officials who must plan for and respond to disasters, and for the public that is deeply affected by the impacts of both the disaster and the pre-event planning efforts. The division between the three components softened as NTHMP scientists and emergency managers worked together to develop the best possible products for the users given the best available science, technology, and planning methods using available funds.

Key words: tsunami mitigation, TsunamiReady, tsunami warning, tsunami evacuation, tsunami planning

Abbreviations: EOC – emergency operations center, FEMA – Federal Emergency Management Agency, NOAA – National Oceanic and Atmospheric Administration, NTHMP – National Tsunami Hazard Mitigation Program, NWS – National Weather Service, RCTWC – Redwood Coast Tsunami Work Group, SEMS – Standardized Emergency Management System, TRC – TsunamiReady Communities, USCG – U.S. Coast Guard, USGS – U.S. Geological Survey, WSLTWG – Washington State/Local Tsunami Work Group

★Author for correspondence: Tel: + 1-425-487-4645; Fax: + 1-425-487-4613; E-mail: chris.jonientztrisler@dhs.gov

1. Background: Tsunami Planning Needs in 1994

In 1994, prior to the start of the NTHMP, the spectrum of tsunami planning activities in at-risk communities in the program states ranged from very little to quite extensive efforts. On 4 October 1994 a Mw 8.3 earthquake in the Kurile Islands triggered a tsunami warning that highlighted this diversity of planning. Often neighboring communities did not show consistent interpretations or responses to tsunami warning messages. In fact, local emergency managers exhibited a range of emotions including confusion, frustration, and anger in reaction to the 1994 event. State and federal emergency managers asked "Why the inconsistencies and turmoil?" and "What can be done to help communities?"

Local emergency managers from eleven communities in Northern California, Oregon, and Washington answered questions during a brief survey after the event (Jonientz-Trisler, 1994). The questions concerned perceived vulnerability and readiness levels, tsunami "safe" locations in the community, the existence of evacuation routes and plans, safe evacuation times, how well local emergency managers understood the 4 October tsunami warning message, and how they responded, including what methods they used to make decisions. Answers indicated that vulnerability and readiness levels varied, responses to the warning varied greatly, and that the warning information system needed improvement. The study recommended ways for federal and state agencies to assist communities to improve vulnerability and readiness levels. Recommendations suggested agencies should develop a regional strategy to provide more consistency in school tsunami plans and drills; make information more timely and usable; have scientists ask responders what kinds of information systems, formats, and tools they require for effective response; and have responders ask scientists what limits exist for information and tools that they base response decisions upon.

Shortly after the 1994 tsunami, NOAA hosted several state/federal agency meetings to develop a strategy to meet the needs of local communities (Tsunami Hazard Mitigation Federal/State Working Group, 1996). West Coast states focused on the need for an improved warning system that gave better and faster information, while Hawaii focused on the need to reduce "false alarms." Meeting participants developed a strategy that includes the following goals: (1) raise awareness of affected populations, (2) supply tsunami evacuation maps, (3) improve tsunami warning systems, and (4) incorporate tsunami planning into state and federal all-hazards mitigation programs.

The National Tsunami Hazard Mitigation Program (NTHMP) was formed in 1996 to implement this strategy. The NTHMP wrote a plan for mitigation projects that would promote the development of "tsunami-resilient communities" (Dengler, 1998). The plan lists five goals that describe the nature of a tsunami-resilient community. Tsunami-resilient communities should: (1) understand the nature of the tsunami hazard, (2) have the tools they need to mitigate the tsunami risk, (3) disseminate information about the tsunami hazard, (4) exchange information with other at-risk areas, and (5) institutionalize planning for a tsunami disaster.

2. Planning Activities that Met the 1994 Needs

A simple plan guided the early years (1996–2001): map the hazard and determine the potential risk level; then inform government officials, residents, and visitors about preparedness, response, and recovery tools such as evacuation brochures, media events, videos, signs, draft legislation, and regulations, and more recently coordination with the TsunamiReady Community Program. NTHMP uses a Tsunami-Resilient Communities Activities Matrix (Table I) to track progress on developing products to meet the goals of the mitigation projects plan. The matrix is broken into planning elements to implement the goals. The Education Planning Element implements both Goal 1 (understanding the nature of the hazard) and Goal 3 (disseminating information about the hazard). Both of the Planning Elements called Tools for Emergency Managers and Building and Land Use Guidance implement Goal 2 (having tools to mitigate the risk). The Information Exchange and Coordination Planning Element implements Goal 4 (exchanging information with other at-risk areas). And the Long-term Tsunami Mitigation Planning Element implements Goal 5 (institutionalize planning for a tsunami disaster). The program uses this information to measure accomplishments and refine goals for future years. The matrix is also a reference to identify existing products.

3. New Strengths Since 1994 and Future Areas of Activity

The first successful accomplishment was the installation of consistent tsunami evacuation signage. Alaska, California, and Washington agreed to adopt Oregon's evacuation sign design (Hawaii already had other signs installed). There is a strong theme of sharing within the NTHMP and time and money is saved by adapting products or pooling resources to develop community products. Other tsunami products, adopted by other states, include educational products such as videos, and information products for targeted audiences like tourists and local officials; tools for emergency managers such as inundation maps, evacuation route brochures, warning programs and guidance, needs assessments and surveys, and some guides for codes, construction, zoning, and land use; information exchange mechanisms like multi-jurisdiction and interdisciplinary workshops and tsunami advisors; and long-term mitigation activities such as all-hazards planning and formal or informal state and local tsunami work groups. Most of these products did not exist in 1994 in the West Coast states. Hawaii and Alaska were an early source of tsunami knowledge for other states but all five states have greatly improved their stock of tsunami mitigation tools since NTHMP's inception. States preferred to develop in-house expertise to produce inundation maps. In order to address issues of consistency in map production the NOAA Tsunami Inundation Mapping Effort (TIME) Center provides scientific and

Table I. Mitigation Strategic Implementation Plan Accomplishments; Tsunami-Resilient Communities Activities Matrix (August 2003)

Planning elements	NTHMP accomplishments	Future directions
Education element – Goal 1: "Understand the risk," Goal 3: "Disseminate risk information"		
Evacuation and Educational Signs	Alaska, California, and Washington adopted Oregon's evacuation sign design. Hawaii had existing signs.	Continue to offer to communities and maintain
Media Materials	Hawaii, Washington report some available	Develop
Public Info Products	All five states have various public information products available	Integrate social science input for successful message to public
Public Service Announcements	Hawaii had existing PSAs, Washington reports some available	Develop with social science input for a successful message to public
Cost/Benefit of Tsunami Mitigation for Businesses	Hawaii is developing a product	Develop
State and Local Videos	All five states have or are developing a tsunami video using local info, including some Native American oral histories	Continue
Curriculum Materials	Hawaii, Washington, and Oregon report available school curriculum	Continue
Library-type Materials	Hawaii and Washington report available library-type materials	Continue
Training Materials	Hawaii and Washington report available training materials	Develop training materials when the need for it is identified

Table I. Continued

Planning elements	NTHMP accomplishments	Future directions
Tsunami Info for Tourists	All five states have tsunami info available for tourists at hotels, restaurants, on the beach, etc.	Integrate social science input for successful message to tourists
Tsunami Info for State and Local	All five states have tsunami info for state and local officials available	Maintain and update
Public Education	All five states have public education materials available	Integrate social science input
Tools for emergency managers element – Goal 2: "Tools to mitigate the risk"		
Inundation Maps	All five states have at least some maps, some have most communities mapped. Obstacles: lack of bathymetry and funds. States that had some inundation maps prior to NTHMP have refined earlier map products based on new technology and modeling methods.	Support bathymetry and funding efforts and partners. Continue to develop maps as bathymetry and funds allow.
Evacuation Routes	All five states have determined at least some evacuation routes with communities	Continue
Evacuation Brochures	Most states have assisted communities in developing evacuation brochures	Continue
Warning Programs	All five states have warning programs	Continue to improve where possible
Local Warning System Guidelines	Hawaii, Oregon, Washington report local warning system guidelines available	Continue development
Guides for Unmapped Communities	Hawaii, Washington report guides for unmapped communities available	Continue development
Community Needs Assessments	All five states have some level of community needs assessments beyond early NTHMP estimates of needs	Continue development with help of social scientists
Surveys	All five states have used tsunami surveys to guide and measure activities	Continue development with help of social scientists

Table I. Continued

Planning elements	NTHMP accomplishments	Future directions
Building and land use guidance element – Goal 2: "Tools to mitigate the risk"		
Codes and Construction Guides	California, Hawaii, Oregon report some available codes and construction guides, Washington reports in development	Continue development
Zoning Regs and Land Use Guides	California, Hawaii, Oregon, Washington report some available zoning regulations and/or land use guides	Continue development
Infrastructure Guides	Hawaii, Washington report some available infrastructure guides	Continue development
Vegetation Guides	Hawaii, Washington report some available vegetation guides	Continue development
Vertical Evac Guides	Hawaii, Washington report some available vertical evacuation guides	Continue development
Information exchange and coordination element – Goal 4: "Exchange information with others"		
Coast Jurisdiction Contact	All five states have contact with their coastal jurisdictions on tsunami planning issues	Continue
Meetings with different disciplines	All five states have fostered meetings between different disciplines that deal with tsunami issues	Continue
Resource Center to catalog products	Hawaii, Washington report available resource center to catalog products	Continue to add materials and to share
Web Page Development	Hawaii, Oregon, Washington report available web site info and offer links to other tsunami web sites	Continue to update
Work with non-NTHMP States	NTHMP is working to exchange information and products with U.S. territories, the Caribbean, Japan, New Zealand through various members	Continue to support and exchange info with others

Table I. Continued

Planning elements	NTHMP accomplishments	Future directions
Tsunami Workshops	All five states have held some workshops focused on a variety of tsunami issues, some multi-state	Continue to explore issues in workshops
Tsunami Technical Advisor Access	All five states plan to use or have used a technical tsunami advisor before and during tsunami events. Hawaii and Alaska pre-existed.	Continue
Long-term tsunami mitigation element – Goal 5: "Institutionalize tsunami planning"		
State/Local Tsunami Work Groups	Most states have state/local tsunami workgroups bringing more than one county together to work issues. This helps reduce staff turnover effects.	Continue
State Tsunami Mitigation Planning	All five states must plan and assist local jurisdictions in planning for tsunami and other hazards	Continue
Incorporate Tsunami into All-Hazards Planning	All five states at risk to tsunami are incorporating it in their all-hazard mitigation plans through the DMA2000 requirements	Continue
Post-Tsunami Recovery Guide	Hawaii reports this in development. The Mitigation Subcommittee also has made this a priority national product to develop	Develop
Loss Estimation	Hawaii reports this in development. The Mitigation Subcommittee also has discussed this as a priority product	Develop
Local Gov't Tsunami Planning Guides	California, Hawaii, Oregon, Washington reports this available or in development.	Develop
Tsunami Legislation	Hawaii, Oregon, Washington report some tsunami legislation available or in development	Develop

technical guidance and assistance to states and developed a preliminary set of best practices. There are plans to archive modeling and mapping products and to establish a formal program for systematic review and improvement of existing inundation and evacuation maps (González *et al.*, this issue).

The NTHMP also develops national level products that require more resources than any one state can afford, but apply to all states. Examples include consistent initial public information products, a guidance document about planning and designing for tsunami hazards (National Tsunami Hazard Mitigation Program, 2001), a guidance document for the public about ways to survive a tsunami (Atwater *et al.*, 1999, 2001), a strategic implementation plan for the mitigation component of the NTHMP (Dengler, 1998), a mechanism for disseminating a broad range of tsunami information to local and congressional officials (National Tsunami Hazard Mitigation Program, 1999–2004), a report to Congress and others on the accomplishments of the Mitigation Subcommittee of the NTHMP (Jonientz-Trisler and Mullin, 1999), and a tsunami warning procedures guidance document (Oregon Emergency Management and Oregon Department of Geology and Mineral Industries, 2001). A project recently funded brings engineers from all five states together to address design of a structure that might withstand both severe ground-shaking and tsunami forces. Future projects under discussion now include a tsunami loss projection study for the five states. The NTHMP provides resources and works with local jurisdictions to develop the most effective products possible. One popular product was modified and translated for use by non-English speakers in this country and in South America (Cisternas and Martínez, translators, 2000). Each state has greatly benefited from the NTHMP accomplishments (see Appendix).

A selective list of some NTHMP mitigation products to promote tsunami-resilient communities include

- Signage
  - tsunami hazard zone signs
  - evacuation signs
  - educational signs
- Evacuation Brochures
  - for homes, visitor centers, and hotels
- Published guidance for
  - surviving a tsunami
  - planning and designing for tsunami hazards
  - warning systems procedures
- A newsletter to disseminate and exchange information on tsunami facts, products, activities, and history
- Public information and outreach products
  - tsunami bookmarks that tell what to do

  - coffee mugs that show what to do
  - trivia puzzles using tsunami facts and words
  - family disaster cards, magnets, stickers, and tent cards
  - tsunami place mats for restaurants
  - coloring books
  - ice scrapers
- School curriculum and booklets for children
- Videos

These products can be acquired through information provided on the NTHMP web site (www.pmel.noaa.gov/tsunami-hazard/).

4. Survey Tools Measure Accomplishments

A May 2001 survey was designed by Dr. Trish Bolton to measure the perceived level of readiness and understanding of tsunami risk by local emergency managers using different questions from the 1994 survey, because there had been no tsunami warning event for the West Coast since 1994. The Bolton survey also assessed the use and perception of NTHMP developed and distributed products supplied to local emergency managers. During the evaluation of the May 2001 survey results, a tsunamigenic earthquake near the coast of Peru triggered Pacific-wide watch/warning messages. This allowed the 1994 survey to be repeated in June 2001 (Jonnientz-Trisler, 2001). The results of the June 2001 survey were compared to those of the October 1994 survey, using the same questions to local emergency managers in much the same communities. Asked whether the tsunami information received during the watches/warnings provided a clear community risk and were timely, updated, understandable, usable, and whether the terminology was clear between "watch" and "warning," local emergency managers responded positively only 36 to 45% of the time in 1994, but responded positively from 79 to 93% of the time in 2001 (Figure 1). Also, the results of the Bolton survey indirectly support many of the survey results from the June 2001 survey. In May most respondents claimed that tsunami readiness was much better. When asked to rate six factors as reasons for this, respondents chose better plans and coordination, better information and public education "What to do" as the top three reasons for the improvement (Table II).

In 2003 following a tsunamigenic earthquake in Japan, a tsunami watch/warning message for Alaska was broadcast providing another opportunity to measure local responder satisfaction level with the warning messages and system. Responses in 2003 to the same six questions asked in 1994 and 2001 elicited the highest levels yet of positive answers for all but one question, and the difference for that question was not statistically significant. The conclusion is that the largest leap in improvement occurred between 1994

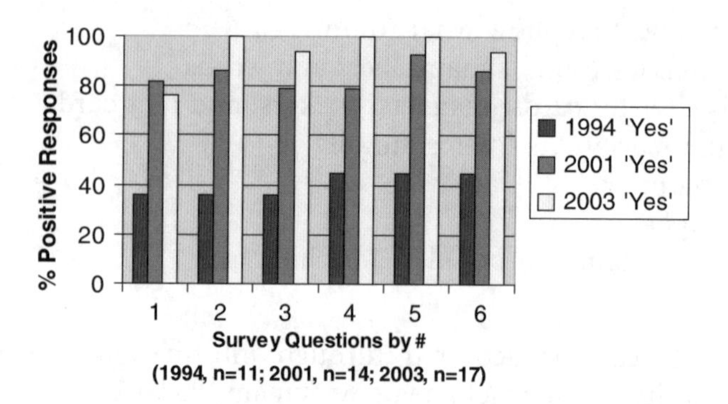

Figure 1. Responses to the questions were dramatically more positive in 2001 than in 1994, indicating that local emergency managers found the improved warning information system much more clear, timely, understandable, and usable than they did in 1994. Answers to the questions in 2003 indicated yet higher levels of satisfaction with the system since the dramatic improvement indicated in 2001. Questions: (1) Based on information provided, was the risk to the community clear to you? (2) Was the information you received on the tsunami timely? (3) Was the information you received on the tsunami updated regularly? (4) Was the information you received on the tsunami understandable? (5) Was the information you received on the tsunami usable? (6) Is present terminology clear regarding "watch" and "warning"?

(36–45% positive response) and 2001 (79–93% positive response), but slightly more improvement (94–100% positive response for all but one question) was measured between 2001 and 2003 overall (Figure 1). NTHMP continues to work with local responders to provide warnings in ways most useful to them.

5. A 5-Year Program Review Measures Accomplishments

During the August 2001 5-year program review, NTHMP members described products and activities and results of three surveys of local

Table II. Survey of 16 local emergency managers showing the factors deemed critical for improvement in their level of tsunami readiness

Factor	Responses (%)
Better plans and coordination	88
Better information	75
Public education: "What to Do"	63
Train responders	38
Better technology	31
Other	19

emergency managers in communities in California, Oregon, and Washington that had been done between 1994 and 2001. Reviewers commented that many of the activities were commendable. They also encouraged more effectiveness in mitigation activities by suggesting the program include social scientists to provide input on how to ensure the right message was being most effectively delivered to users of the information. The program addressed these suggestions by adjusting future goals to incorporate social scientists' input on program activities currently underway (Bernard, this issue).

6. Incorporation of Other Programs and Partners and Plans for the Future

Internally, the NTHMP collaboration among scientists and emergency managers has grown and will continue to do so. The value of working together by interweaving all aspects of the program, such as hazard identification, modeling, mapping, community outreach, evaluation, and planning, is clear. Scientists and emergency managers commonly attend one another's topical meetings and provide input on activities rather than work only within one's specific discipline. Mitigation focuses on the translation of the science and technology into user-friendly planning and education products for federal, state, and local officials who must plan for and respond to disasters, and for the public that is deeply affected by the impacts of both the disaster and the pre-event planning efforts.

Externally, as the 5-year program review suggested, the program will collaborate with other programs and disciplines. One of the successes of the NTHMP has been the collaboration with a National Weather Service (NWS) program, the TsunamiReady Communities (TRC) Program (http://wcatwc.arh.noaa.gov/tsunamiready/tready.htm). NWS worked with NTHMP members to design a program based on the StormReady Communities Program. A community must meet certain criteria to be designated a TsunamiReady Community and must continue to meet renewed certification standards in order to keep that designation. These criteria include

- An Emergency Operations Center
- The ability to disseminate a tsunami warning (sirens, local media)
- A tsunami hazard plan
- A community awareness program
- Multiple ways to receive NWS tsunami warnings
  1. Emergency Management Weather Information Network (EMWIN) receiver
  2. NOAA Weather Radio (NWR)
  3. NOAA Weather Wire drop

Currently there are 15 communities designated TsunamiReady, including one tribal nation (Ocean Shores, Long Beach, and the Quinault Indian Nation, WA; Cannon Beach, OR; Homer, Sitka, Seward, and Kodiak, AK; and Crescent City, CA).

NTHMP is seeking the use of social science research to effectively measure the success of planning and education products and to be able to modify them to increase their effectiveness. Initially local emergency managers responded to surveys and attended workshops designed to find out from them what warning messages and protocols were working well and what needed improvement to better serve their needs, for example, warning message format or training in procedures. NTHMP members also researched and compiled a guidance document for local responders describing existing systems, equipment, protocols, and procedures, and their pros and cons (Oregon Emergency Management and the Oregon Department of Geology and Mineral Industries, 2001). Currently, members are working to incorporate tsunamis in several all-hazards programs in the western states, including FEMA's Pre-Disaster Mitigation Grant all-hazards plans required for states and communities. Members are also working with the National Flood Insurance Program Community Rating Service staff to provide input on reasonable credits for tsunami activities, and this will be an incentive for coastal communities to address both flood and tsunami hazards. Finally, members are working toward incorporating tsunami hazard into the existing disaster response and recovery system through providing technical advice and information, and forming some more formal liaison process that can be used shortly after a tsunami disaster occurs.

We began the NTHMP with a vision of helping build "tsunami-resistant communities," but based on the expanding toolbox the NTHMP is developing over the years, the more realistic vision for the program has become "tsunami-resilient communities." This word change does not reflect a change in the goals described in the strategy envisioned in 1998 (Dengler, 1998). We have communities that, short of being picked up and relocated elsewhere, will not be able to oppose the forces of a tsunami that resistance implies. A tsunami-resilient community is one that will take advantage of actions, products, and policies that can help it bounce back from the inevitable tsunami event that will surely come out of the near or far future. A tsunami-resilient community may suffer some inevitable damage, but will have planned, exercised, and educated its citizens and its leaders in ways to save lives, protected as much property as possible, tried to ensure safe locations for critical functions the community needs, and will use lessons from a tsunami event suffered by their community or other communities to improve their level of resilience for future events.

Appendix

A. ALASKA

The state of Alaska has benefited from several NTHMP product developments. Information obtained from tsunameters (González *et al.*, 2003) allows tsunami warnings/watches to be disseminated more accurately to tsunami-prone communities. Through the state's Tsunami Inundation Mapping Program, tsunami inundation maps for communities along the Gulf of Alaska are being generated. Inundation maps for Kodiak City, U.S. Coast Guard (USCG) station, and Women's Bay are complete; maps for Homer and Seldovia are in progress; Sitka and Seward and other communities will be mapped in the future.

The TsunamiReady Community program promotes tsunami hazard preparedness by supporting better and more consistent tsunami awareness and mitigation efforts. The main goal is improvement of public safety during tsunami emergencies. The communities of Seward, Homer, Sitka, and Kodiak are certified "TsunamiReady," and the Borough and City of Kodiak have nearly completed requirements to become TsunamiReady.

The Tsunami Sign Program is a joint NTHMP effort to coordinate and disseminate consistent tsunami information. Alaska contacted all coastal communities at risk to locally generated or distant tsunamis and offered standardized tsunami hazard signs. Signs are now installed in Sitka, Sand Point, Seward, Kodiak, and Homer. Also, the Alaska Department of Parks and Recreation installed signs in Shoup Bay, a remote area inundated to as much as 170 feet above sea level in 1964 and now frequented by hikers and kayakers (Lander and Lockridge, 1989).

Tsunami hazard awareness, education, and outreach are a priority for Alaska. Numerous materials were produced and distributed to communities, businesses, and the public, including school curriculum, coloring books, bookmarks, emergency contact cards, magnets, tent cards, ice scrapers, and decals. Brochures are produced for TsunamiReady Communities and include tsunami information, evacuation route maps, shelter locations, NOAA Weather Radio information, and survival/safety tips.

In conjunction with a "Quake Cottage" program, tsunami preparedness is presented to those communities where a tsunami hazard exists. The "Quake Cottage" is a small van equipped with a shake table that simulates an earthquake. The public can experience the ground shaking associated with a large earthquake in a safe environment. The cottage has been present at many large community events such as Alaska State Fair, the Kodiak Crab Festival, the Ninilchik Fair, the Kenai River Festival, and the Governor's Picnic.

B. CALIFORNIA

California efforts have concentrated on creating the knowledge base essential for building a constituency to support tsunami-planning efforts. This involved creation of a coalition of emergency management representatives of coastal counties, state agencies responsible for regulating development, coastal parks, transportation, and geological mapping. This effort produced a consensus strategic plan for allocating funds for mapping, mitigation planning, guidance development, and the initiation of evacuation planning. The priorities of this State Tsunami Steering Committee have been to complete inundation projections for the 500+ mile coastline, emphasizing the highly populated areas of southern and central California, followed by the less populated coastal areas north of the San Francisco Bay Region.

The availability of local inundation projections and maps by local governments fosters interest in mitigation at both local and state levels. In San Diego, Santa Barbara, Los Angeles, San Mateo, and San Francisco counties, local evacuation planning efforts were initiated in 2000 and 2001. Unfortunately, the events of 11 September 2001 have shifted local priorities and delayed further implementation of local planning. At the state level, the availability of maps is drawing the interest of the California Geological Survey's mandated Hazard Mapping Program. The State's Hazard Mapping Program will address tsunami inundation when recurrence and probabilities of occurrence can be established, consistent with California's earthquake, flood, landslide, and liquefaction risk assessment programs.

In order to ensure a consistent hazard identification and response planning processes among the coastal counties, the California Office of Emergency Services developed and made available a *Local Planning Guidance* to integrate tsunami efforts with the multi-hazard mitigation and preparedness procedures of the State's Standardized Emergency Management System (SEMS).

California served as the project manager for the development of *Designing for Tsunamis*, a guidance document for land use planners and local government development decision makers. The publication provides examples of planning, site development, and building configuration approaches that mitigate the impacts of tsunami inundation.

In the northern counties of Del Norte, Humboldt, and Mendocino, the Redwood Coast Tsunami Work Group (RCTWC) coordinates education and preparedness efforts among local and state agencies.

Similar organizing efforts in southern California have been delayed since the fall of 2001 by the priority placed at federal, state, and local levels of government resulting from the threat of weapons of mass destruction and terrorism (WMD/T) events.

C. HAWAII

The state of Hawaii has directly benefited from several program accomplishments. Operational deployment of six NOAA tsunameters off the Alaska Peninsula and Aleutian Islands, off Pacific Northwest coasts, and in the eastern equatorial Pacific has been successfully accomplished to more accurately evaluate tsunamis approaching Hawaii and other U.S. coasts from afar. Moreover, the government of Chile has purchased one additional tsunameter to enhance the Chile warning system. NOAA will deploy this tsunameter off the Chilean coastline in the fall of 2003. This tsunameter will provide Hawaii with a timely and accurate measure of a Chilean tsunami (e.g., the 1960 Pacific wide destructive tsunami), and is hopefully the first in an internationally supported network of tsunameters that will share vital deep ocean data among all nations affected by tsunamis.

Implementation of operational NOAA tsunami wave forecast is now underway. Prior to the NTHMP, the only NOAA forecast product was time of tsunami wave arrival. Now with a tsunameter network and coastal instruments, tsunami wave forecasts are possible. Such forecasts are essential to reduce "false alarm" tsunami evacuations in Hawaii and all the other Pacific states.

NTHMP upgraded parts of the U.S. Geological Survey's seismic network and facilitated the use of these data at both the West Coast/Alaska and Pacific Tsunami Warning Centers to more rapidly locate and accurately measure earthquake magnitude and tsunami generation potential.

NTHMP has made possible numerous Hawaii-based tsunami scientific, mitigation, and public awareness initiatives (e.g., distant and local tsunami forecast and shoreline wave inundation models, installation of coastal inundation detectors on the island of Hawaii to rapidly detect locally generated tsunamis, upgraded Civil Defense emergency response capabilities, April Tsunami Awareness Month, media training workshops, public safety videos, etc.).

D. OREGON

In Oregon the focus has been on education, inundation, and evacuation maps, signs, workshops, guidance documents, and legislation. Oregon produced the following educational products: a tsunami video showcasing the Oregon tsunami hazard, grade 7–12 tsunami school curriculum, brochures, and a variety of other materials such as tent cards, stickers, magnets, and bookmarks.

The state produced detailed tsunami inundation maps for six coastal areas. Prior to the detailed maps, simple tsunami inundation maps were developed for the entire coastline as part of legislation (Oregon Senate Bill

379) that limits construction of new critical (e.g., fire stations) and essential (e.g., schools) facilities in the officially designated tsunami inundation zone. Seventeen GIS-based tsunami evacuation maps covering 25 communities were produced using the latest inundation estimates calculated by accepted tsunami modeling methods. The format is consistent and locals have input in designating evacuation routes and format for user-friendly public use of the maps. Prior to these GIS based maps, evacuation maps had been created by local jurisdictions with and without financial assistance from the National Tsunami Hazard Mitigation Program. The goal is to have these evacuation maps for all areas on the coast.

Tsunami signs produced for Oregon include hazard zone, entering and leaving hazard zone, evacuation route, and evacuation site signs. They were distributed and installed in many locations on the coast. A sign installation guidance document was produced to assist in their placement. Interpretive signs, that include tsunami science and tsunami response information, were installed at a number of locations. Oregon is in the process of producing a historical marker sign for Siletz Bay that illustrates the probable impact of the 1700 Cascadia tsunami on a native village.

Three workshops were held: a general tsunami workshop in 1998, a lodging facility planning workshop in 2000, and a tsunami warning workshop in 2002. In addition to the guidance documents mentioned above, a lodging facility planning guidance document and a guide explaining the procedures for compliance with Oregon Senate Bill 379 were also developed. Additional legislation requires schools in the inundation zone to conduct tsunami evacuation drills as well as earthquake drills. Legislation was introduced in the 2003 state legislative session that would require lodging facilities to post tsunami information.

Oregon's tsunami hazard, interpretation, and evacuation signs were used as a model for similar signs in Washington, California, and Alaska to provide consistency for the public who live and travel along the North and West Pacific Coast. Many of Oregon's educational products (brochures, tent cards, stickers, magnets, and bookmarks) were also modified for use in the other states. And although Oregon developed many of its products prior to the NTHMP, the NTHMP funds and accomplishments have allowed Oregon to refine and expand its list of products and activities.

E. WASHINGTON

The state of Washington used the NTHMP Federal/State model to develop its state tsunami mitigation program at a more local level and is guided by the Washington State/Local Tsunami Work Group (WSLTWG). This group recommends priority areas of focus and provides input and active involvement. The group is key in translating the science and technology into usable information for the public and local officials. The State/Local Tsunami Work Group has developed tsunami brochures to provide information on the tsunami hazard. These brochures include evacuation maps, NOAA Weather Radio information, and tsunami safety tips. In recent years some areas of emphasis include making the warning system more efficient and measuring the effectiveness of tsunami program activities and products for the public (see paper in this edition by George Crawford for more details).

Recently, the WSLTWG adopted the NOAA Weather Radio "All-Hazards" Warning System to warn citizens quickly and effectively of not only tsunami hazards but also other natural or man-made hazards. To implement the NWR strategy and address a gap in warning coverage, the group developed a partnership to add a repeater to the NWR system that provides complete coverage to the coast of Washington and to shipping lanes off the coast. Also, they developed a new notification system to disseminate time-critical tsunami hazard information to the public on beaches and in high-traffic areas. These innovative developments and processes gave rise to the Tsunami Warning/Evacuation Cycle that was also developed. In concert with an array of deep ocean tsunameters, land-based seismic sensors, and warning messages issued by the tsunami warning centers, the NWR provides a means to expeditiously get critical decision-making information to emergency managers, elected officials, and first responders.

The state also examined residents' and visitors' perceptions of the tsunami hazard by working with David Johnston, from the Institute of Geological and Nuclear Sciences in New Zealand, who is experienced with hazard perception surveys. An element of the survey focused on the public's understanding and knowledge of how a tsunami warning is received and disseminated to them and their preparedness to deal with this hazard. One of the findings concluded that approximately half of all students were unaware of the elements of the state's tsunami warning system or who is responsible for issuing the warning. As a result of Johnston's study the booklet "How the Smart Family Survived a Tsunami" (elementary edition – K–6) was revised. The booklet now addresses the tsunami warning process, the Washington Tsunami Alert and Notification System, and actions people should take when a tsunami warning is received. It also has information on a family disaster plan and disaster supply kit.

With September designated as Weather Radio Awareness month in Washington, the Work Group's goal is to have NOAA Weather Radios become as common as smoke detectors in homes and businesses statewide to help protect lives and property from natural and technological hazards.

Acknowledgements

We would like to thank Brian Atwater of the U.S. Geological Survey, Eddie Bernard of the National Oceanic and Atmospheric Administration, and Lt. Allan Yelvington of the U.S. Coast Guard for sharing their insights and providing valuable commentary.

Reference

Atwater, B. F., Cisternas, V. M., Bourgeois, J., Dudley, W. C., Hendley II, J. W., and Stauffer, P. H.: 1999, Surviving a Tsunami – Lessons from Chile, Hawaii, and Japan. USGS Circular 1187, 18 pp.

Atwater, B. F., Cisternas, V. M., Bourgeois, J., Dudley, W. C., Hendley II, J. W., and Stauffer, P. H.: 2001, Sobreviviendo a un tsunami: lecciones de Chile, Hawai y Japòn. U.S. Geological Survey Circular 1218, 18 pp.

Bernard, E. N.: 2005, The U.S. National Tsunami Mitigation Program: A Successful State-Federal Partnership. *Nat. Hazards* **35**, 5–24 (this issue).

Cisternas, M., and Martínez, P. (translators): 2000, Como sobrevivir a un maremoto: 11 lecciones del tsunami occurido en el sur de Chile el 22 de mayo de 1960. Servicio Hidrográfico y Oceanográfico de la Armada de Chile, 14 pp.

Dengler, L.: 1998, Strategic Implementation Plan for Tsunami Mitigation Projects, approved by the Mitigation Subcommittee of the National Tsunami Hazard Mitigation Program, April 14, 1998. Technical Report NOAA Tech. Memo. ERL PMEL-113 (PB99-115552), NOAA/Pacific Marine Environmental Laboratory, Seattle, WA. http://www.pmel.noaa.gov/pubs/PDF/deng2030/deng2030.pdf.

González, F. I., Sherrod, B. L., Atwater, B. F., Frankel, A. P., Palmer, S. P., Holmes, M. L., Karlin, B. E., Jaffe, B. E., Titov, V. V., Mofjeld, H. O., and Venturato, A. J.: 2003, Puget Sound Tsunami Sources – 2002 Workshop Report, A contribution to the Inundation Mapping Project of the U.S. National Tsunami Hazard Mitigation Program. NOAA OAR Special Report, NOAA/OAR/PMEL, 34 pp.

González, F. I., Titov, V. V., Mofjeld, H. O., Venturato, A., Simmons, S., Hansen, R., Combellick, R., Eisner, R., Hoirup, D., Yanagi, B., Young, S., Darienzo, M., Priest, G., Crawford, G., and Walsh, T.: 2005, Progress in NTHMP Hazard Assessment. *Nat. Hazards* **35**, 89–110 (this issue).

Jonientz-Trisler, C.: 1994, Cascadia Response to October 4, 1994 Kurile Islands Mw 8.3 Earthquake-Induced Tsunami Warning. Abstract, American Geophysical Union, Fall 1994 Meeting, Special Session on the Kurile Island Earthquake and Tsunami.

Jonientz-Trisler, C.: 2001, The Mitigation Strategic Implementation Plan: Toward Tsunami Resistant Communities. In: *Proceedings of the International Tsunami Symposium 2001 (ITS 2001)* (on CD-ROM), NTHMP Review Session, R-8, Seattle, WA, 7–10 August 2001, pp. 119–159. http://www.pmel.noaa.gov/its2001/.

Jonientz-Trisler, C., and Mullin, J.: 1999, 1997–1999 Activities of the Tsunami Mitigation Subcommittee: A Report to the Steering Committee NTHMP. FEMA Region 10 publication, 45 pp., Appendices.

Lander, J. F., and Lockridge, P. A.: 1989, *United States Tsunamis (Including United States Possessions) 1690–1988*. NOAA Publication 41–2, 265 pp.

National Tsunami Hazard Mitigation Program: 1999–2004, *TsuInfo Alert*. Prepared for the NTHMP by C. Manson and L. Walkling, Washington State Department of Natural

Resources, Division of Geology and Earth Resources, published bi-monthly as a newsletter.

National Tsunami Hazard Mitigation Program: 2001, *Designing for Tsunamis*, 59 pp.

Oregon Emergency Management and Oregon Department of Geology and Mineral Industries: 2001, *Tsunami Warning Systems and Procedures: Guidance for Local Officials*. Oregon Department of Geology and Mineral Industries Special Paper 35.

Tsunami Hazard Mitigation Federal/State Working Group: 1996, Tsunami Hazard Mitigation Implementation Plan – A Report to the Senate Appropriations Committee, 22 pp., Appendices, http://www.pmel.noaa.gov/tsunami-hazard/hazard3.pdf.

Natural Hazards (2005) 35: 141–153

The Role of Education in the National Tsunami Hazard Mitigation Program

LORI DENGLER
Department of Geology, Humboldt State University, #1 Harpst St., Arcata, CA 95521, USA (Tel: +1-707-826-3115; Fax: +1-707-826-5241; E-mail: ladl@humboldt.edu)

(Received: 24 September 2003; accepted: 6 April 2004)

Abstract. Tsunami education activities, materials, and programs are recognized by the National Tsunami Hazard Mitigation Program (NTHMP) as the essential tool for near-source tsunami mitigation. Prior to the NTHMP, there were no state tsunami education programs outside of Hawaii and few earthquake education materials included tsunami hazards. In the first year of the NTHMP, a Strategic Plan was developed providing the framework for mitigation projects in the program. The Strategic Plan identifies education as the first of five mitigation strategic planning areas and targets a number of user groups, including schools, businesses, tourists, seasonal workers, planners, government officials, and the general public. In the 6 years of the NTHMP tsunami education programs have been developed in all five Pacific States and include print, electronic and video/film products, curriculum, signage, fairs and workshops, and public service announcements. Multi-state education projects supported by the NTHMP include TsuInfo, a bi-monthly newsletter, and *Surviving a Tsunami*, a booklet illustrating lessons from the 1960 Chilean tsunami. An additional education component is provided by the Public Affairs Working Group (PAWG) that promotes media coverage of tsunamis and the NTHMP. Assessment surveys conducted in Oregon, Washington, and Northern California show an increase in tsunami awareness and recognition of tsunami hazards among the general population since the NTHMP inception.

Key words: mitigation, evacuation planning, education, awareness, near-source tsunami

Abbreviations: CSZ – Cascadia Subduction Zone, FEMA – Federal Emergency Management Agency, NOAA – National Oceanic and Atmospheric Administration, NTHMP – National Tsunami Hazard Mitigation Program, NWS – National Weather Service, PAWG – Public Affairs Working Group, WSSPC – Western States Seismic Policy Council

1. Introduction

From its inception, the National Tsunami Hazard Mitigation Program (NTHMP) has recognized education as a major part of reducing vulnerability to tsunami hazards in the United States. The NTHMP had its origins with the 1992 7.1 (Mw) Cape Mendocino earthquake in Humboldt County, California that produced not only shaking damage, but a small tsunami (González and Bernard, 1993) that was recorded in Northern California, Southern Oregon, and in Hawaii. The location and orientation of rupture strongly suggested an origin on or near the Cascadia subduction zone (CSZ) (Oppenheimer *et al.*, 1993), confirming the capability of the CSZ to produce strong earthquakes and local tsunamis. The tsunami raised the concerns of State and Federal agencies responsible for disaster planning that near-source tsunami hazards were not adequately addressed by the U.S. tsunami warning system. Oregon's Senator Hatfield convened Senate hearings to assess the tsunami vulnerability of the west coast of the United States.

As a result of those hearings, the National Oceanic and Atmospheric Administration (NOAA), the federal agency responsible for issuing tsunami warnings, was tasked with assessing U.S. tsunami vulnerability. A series of workshops on warning systems, tsunami modeling, and education were held to assess existing projects and programs and define mitigation needs (Bernard, 1998). The Education Workshop (Good, 1995) identified two major gaps: a lack of awareness along the west coast of a local tsunami hazard, and confusion among emergency managers and the public about the tsunami warning system.

The workshops formed the basis of the Tsunami Hazard Mitigation Implementation Plan (Tsunami Hazard Mitigation Federal/State Working Group, 1996). The plan defined education as the primary tool for reducing losses from a locally generated tsunami and it identified three primary education needs of people and communities in potentially hazardous zones:

- Recognizing the signs of an impending tsunami
- Understanding what areas are at risk
- Knowing how and when to evacuate

This paper reviews (1) educational efforts of the NTHMP and (2) assessments of the effectiveness of those efforts.

2. The Strategic Plan for Tsunami Mitigation

Mitigation activities of the NTHMP are directed by the Mitigation Subcommittee consisting of two representatives from each state – an emergency manager and a geoscientist – and chaired by the Federal Emergency Management Agency (FEMA) program representative. During the first year of the NTHMP, a Strategic Implementation Plan was developed to assess existing mitigation programs and materials, formulate mitigation strategies, and set priorities for projects (Dengler, 1998). The plan recognizes the different tsunami exposure and unique demographic situations of the five Pacific states and the need to incorporate tsunami efforts into existing earthquake and all-hazard mitigation programs. The goal of the plan is to encourage "tsunami resilient" coastal communities. A tsunami resilient community:

1. *Understands the nature of the tsunami hazard.* Knows the risk that tsunami waves, from both near and far sources, pose to its coastal areas.
2. *Has the necessary tools to mitigate the tsunami risk.* Has defined needed mitigation products and knows how to access and use them.
3. *Disseminates information about the tsunami hazard.* Has identified vulnerable populations, has materials defining areas at risk and safety, evacuation routes, appropriate response, and has developed a dissemination plan to provide information to all users of the coastal area.
4. *Exchanges information with other at-risk areas.* Shares mitigation products with other coastal communities and incorporates lessons from all-hazard mitigation programs into tsunami hazard reduction efforts.
5. *Institutionalizes planning for a tsunami disaster.* Has incorporated tsunami hazard mitigation elements into their long-term all-hazard management plans and has developed a structure to develop and maintain the support of local populations and decision makers for mitigation efforts.

Education is the first of five strategic planning elements. The plan recognizes that technological advances and warning systems cannot protect coastal populations from a near-source tsunami because the first waves may reach the coast within minutes of the event. Local populations must be able to recognize the signs of an impending tsunami and take appropriate action immediately without official direction.

The plan identifies a number of programmatic educational needs, including the nature of the hazard, evacuation information, curriculum, information geared for tourists and other occasional visitors to the coastal zone, and information targeted at decision makers to sustain tsunami mitigation efforts. The plan supports efforts to develop comprehensive educational programs for the diverse users of the coastal environment.

3. State Educational Products

The Strategic Plan defines twelve education planning elements, including print, electronic, audio and video materials, posters and signs, curriculum programs, museums and information centers, public relations efforts, workshops, and other public forums targeted for a variety of audiences. All five Pacific states have developed materials and/or programs that address most of these elements. Education products of the NTHMP are summarized by Jonientz-Trisler and Mullin (1999) and Jonientz-Trisler *et al.* (this issue).

3.1. GENERAL INFORMATION MATERIALS

All five states have developed a variety of print materials on the general nature of the tsunami hazard. These include discussions of near-source versus distant-source events, the multiple wave nature of tsunamis, and the need to head inland or to higher ground immediately after a strong coastal earthquake. They also include information specific to a state or region such as the Cascadia subduction zone in the Pacific Northwest, and the landslide tsunami source in Hawaii. Many of these publications can be accessed on-line at http://www.wsspc.org/tsunami/tsunamipubs.html. In addition to print materials, a variety of educational products were developed by the Oregon Department of Geology and Mineral Industries, including bookmarks, magnets, stickers, and a heat-sensitive tsunami mug, and made available to the other Pacific states and to the general public.

3.2. CURRICULUM

Prior to the NTHMP, tsunamis were touched on only briefly in the FEMA earthquake curricula for K–6 and 7–12 schools. Several activities were developed by Dr. Dan Walker of the University of Hawaii and used in Hawaii, but there were no comprehensive tsunami curricula elsewhere in the United States. Early in the NTHMP, Oregon completed two tsunami curriculum packages, one directed at grades K–6 and the other for junior high and high school. Passage of Oregon State Senate Bill 378 required all Oregon schools in potential inundation zones to teach students in grades kindergarten through eighth grade about tsunamis and practice evacuation drills. The Oregon school program was recognized by the Western States Seismic Policy Council (WSSPC) with an Award in Excellence for outreach to schools in 1999. The Oregon curriculum was revised and adapted by the State of Washington and published in a two-booklet set (Washington Military Department, Emergency Management Division, 2001a, b). The Washington curriculum received the WSSPC Award in Excellence for school outreach in 2001. Tsunami curriculum materials have also been developed at Humboldt State University (HSU) by the Geology Department. Two professional development courses for teachers and student teachers (Geology 700, *Tsunami!*, and *Tsunamis on the North Coast*) are offered by the HSU Geology Department each year and the curricula are currently being adapted for electronic dissemination.

3.3. EVACUATION MAPS AND SIGNAGE

The first action of the NTHMP mitigation subcommittee was to endorse tsunami hazard and evacuation signage developed by the Oregon Department of Transportation (Figure 1). Signs serve an essential educational role by raising community awareness before a tsunami and by notifying people of appropriate evacuation routes for use during a tsunami. Installation of signs has generated media attention that reaches an even larger audience. Washington and Oregon have installed signs in all coastal communities for which inundation maps are available. Oregon maps are posted online at: http://www.oregongeology.com/earthquakes/Coastal/Tsumaps.HTM.

Figure 1. Tsunami evacuation sign, Crescent City, California.

In addition to signs, both states have also developed brochures and other print material to disseminate evacuation information to most coastal communities (Figure 2). Oregon's coastal sign and community education program was first recognized by the WSSPC with an Award in Excellence for outstanding outreach to the general public in 1996. The evacuation map program received a second award in 2003. Tsunami signs have also been posted in Kodiak, Alaska and in Crescent City, California and evacuation guidance materials developed for some communities in each state. Evacuation maps had been completed for all Hawaii coastal cities prior to the NTHMP and maps are published in telephone books.

3.4. PUBLIC SERVICE ANNOUNCEMENTS AND VIDEO PRODUCTS

Oregon, Washington, Northern California, and Hawaii have developed video products related to tsunami hazards. Oregon has produced several videos including public service announcements, a general information video short,

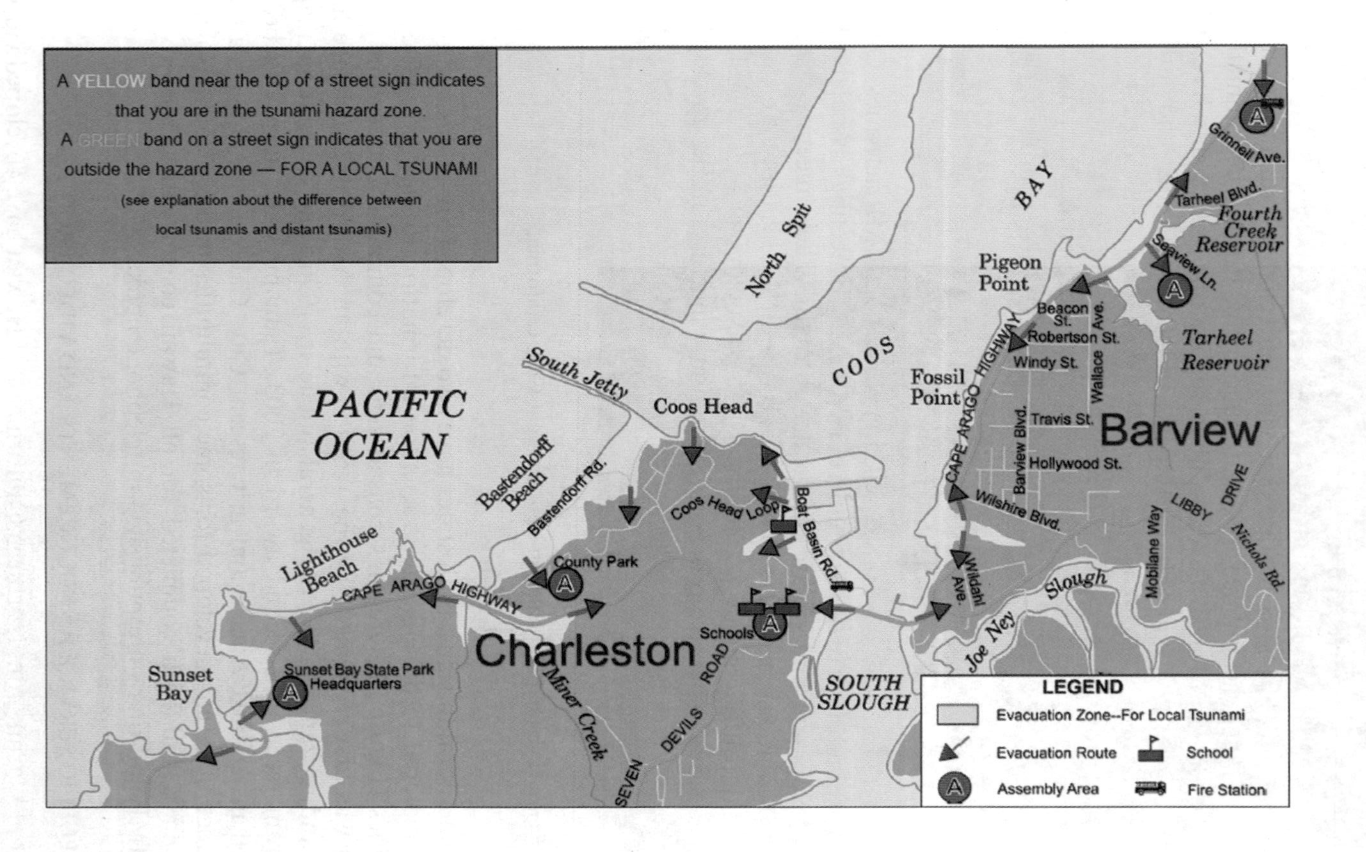

Figure 2. Tsunami evacuation map, Charleston-Sunset Bay, Oregon. Map developed by the Oregon Department of Mining and Mineral Industries.

and "Tsunami: Surviving the Killer Wave" which is part of the 7–12 grade curriculum. Washington has produced public service announcements. Hawaii produced a half-hour tsunami video "Tsunami, Waves of Destruction" that received a 2003 WSSPC Award in Excellence for outreach to the general public. Hawaii has also produced a video directed at tsunami safety for surfers and includes public service announcements during their annual Tsunami Awareness Month. Several public service announcements have been produced by the Humboldt Earthquake Education Center and aired on California's North Coast. These announcements were produced as a collaborative project with local public schools. The project was awarded a WSSPC Award in Excellence for outreach to schools in 1998. (Visit www.pmel.noaa.gov/tsunami-hazards for access to these products.)

3.5. MUSEUMS, FAIRS, WORKSHOPS, AND OTHER FORUMS

All five Pacific States have held workshops or other forums to develop and/or disseminate information to the public as part of NTHMP activities. Oregon and Washington have hosted numerous community meetings associated with the development and release of evacuation maps. These meetings have allowed community members to make decisions on the location of evacuation lines and routes and to get feedback from tsunami experts. Alaska has encouraged an extensive process of community meetings in the development

Figure 3. Tsunami Room at the 2003 Humboldt County Fair, Ferndale, California.

of evacuation maps and has included tsunamis in their Quake Cottage (earthquake simulator) outreach program at fairs and schools. Alaska's Quake Cottage program received a 2003 WSSPC Award in Excellence for outreach to the general public. Hawaii hosts an annual tsunami preparedness month in April of each year with a variety of activities, many at the Pacific Tsunami Museum in Hilo. The Redwood Coast Tsunami Work Group in Northern California has sponsored an Earthquake/Tsunami Education Room at County Fairs for the past 5 years (Figure 3).

4. Multistate Education Products

Each year of the NTHMP has included funding for multi-state projects that benefit more than one state. Two of the funded projects directly addressed education needs.

4.1. TSUINFO PROGRAM

The TsuInfo Program began in 1998 to facilitate information exchange among the member states. The program's primary function is to provide information to emergency managers through a bi-monthly publication that includes news about the NTHMP and other tsunami mitigation activities, tsunami publications, and articles about tsunamis. The newsletter now reaches a much wider audience through pdf versions available on the web at http://www.dnr.wa.gov/geology/tsuinfo/. Recent issues have included articles about great tsunamis of the past such as Lisbon, Portugal (1755) and Krakatoa, Indonesia (1883), tsunami curriculum materials, NOAA Weather Radio, and state tsunami program activities.

4.2. SURVIVING A TSUNAMI: LESSONS FROM CHILE, HAWAII, AND JAPAN

This publication in both English and Spanish (Atwater *et al.*, 1999, 2001) illustrates the experiences of the 1960 Pacific-wide tsunami from the perspectives of Chile, Hawaii, and Japan. Interviews with survivors of the tsunami are organized into a number of lessons: Head for High Ground and Stay There, Expect Many Waves, Climb a Tree, etc. The publication has been included in the Humboldt State University Tsunami Curriculum. It can be downloaded from http://pubs.usgs.gov/circ/c1187/.

5. TsunamiReady Program

The TsunamiReady Program was developed by NOAA's National Weather Service (NWS) in 2001 to promote community tsunami preparedness. It is modeled on the NWS Storm Ready Program and was developed in coordination with and endorsed by the NTHMP Mitigation Subcommittee. To achieve TsunamiReady certification, a community must meet a number of criteria related both to emergency planning/operations and education. TsunamiReady communities must develop a community tsunami awareness program that increases public awareness and understanding of the tsunami hazard. The program encourages consistency in educational materials by setting a single standard for all U.S. coastal communities. Regional NWS offices provide assistance to communities in developing TsunamiReady applications and in maintaining community programs. Fifteen communities/areas have received certification including four each in Oregon and Washington, three in Alaska and two in California and Hawaii. Information about the TsunamiReady Program is posted at http://wcatwc.gov/tsunamiready/tready.htm.

6. Public Affairs Working Group

From its beginning, the NTHMP has recognized the significant role of the media in creating public interest and disseminating information about tsunamis. To facilitate accurate media stories on tsunamis, a Public Affairs Working Group (PAWG) was formed to develop media materials about both the NTHMP and tsunami hazards and facilitate media access to tsunami experts. An example of the effectiveness of the PAWG and NTHMP interaction with the media is provided by the 1998 Papua New Guinea tsunami.

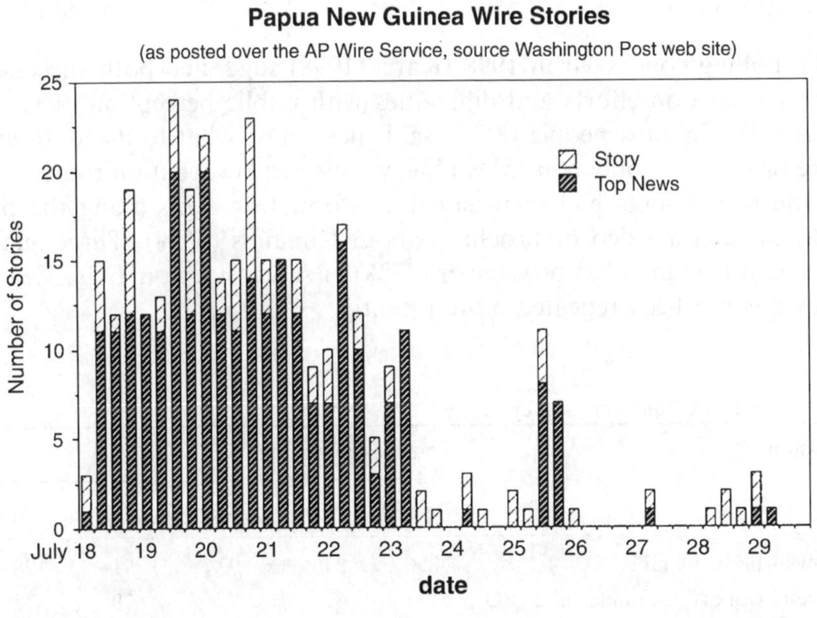

Figure 4. Associated Press Wire Service Stories posted for the Papua New Guinea tsunami, July 1998.

The 1998 tsunami struck the north coast of Papua New Guinea with wave heights of 30–45 feet, killing over 2,000 people. It received an unusual amount of attention from international media. It played particularly large in the United States, Australia, New Zealand, and Japan. The Associated Press listed over 300 postings about the tsunami between July 18 and July 23, making it the leading story of the week (Figure 4). NTHMP Steering Committee members and staff of the Tsunami Warning Centers gave 92 interviews to the media during the month following the tsunami, including all major broadcasting companies (Public Affairs Working Group (PAWG), 1998). Media reports covered not only the disaster itself, but also discussed tsunami hazards in their own regions. Background materials provided by PAWG and the coordination of interviews created higher quality stories with consistent tsunami hazard mitigation themes. Media resources and PAWG reports are posted at: http://www.pmel.noaa.gov/tsunami-hazard/mediaresources.htm.

7. Assessment

A coordinated five-state assessment of the NTHMP education component has not been carried out. However, three in-state studies have addressed the effectiveness of these programs.

7.1. OREGON

Public polling conducted in 1998 (Karel, 1998) suggested both successes in Oregon education efforts and difficulties with public perception of tsunami hazards. While most people (85%) said they know what to do in an earthquake or tsunami, only half (51%) knew their local evacuation route. About half the respondents had seen tsunami information signs along the beach (45%) or seen a video or brochure about tsunamis (55%). Three-quarters read about tsunamis in a newspaper (77%) or saw a story on TV (75%). The survey has not been repeated more recently.

Table I. Northern California survey results

Question	Percent responding "Yes"				
	Apr 93	Nov 93	Mar 95	Jan 96	Apr 01
Knows what tsunami is	78	84	92	91	98
Tsunami can arrive minutes after EQ	51	62	75	70	73
Not safe after 1st wave retreats	65	73	75	81	87
Knows what Cascadia S.Z. is	16	20	29	32	42

Table II. Effectiveness of "Living on Shaky Ground"

Question	Percent responding "Yes"					
	1993		1995		1996	
	No	Yes	No	Yes	No	Yes
Knows what tsunami is	80	91	87	95	85	95
Tsunami can arrive minutes after EQ	60	66	72	77	75	78
Knows what Cascadia S.Z. is	14	33	23	41	19	41

No columns: respondents who had not seen magazine.
Yes columns: respondents who had seen magazine.

7.2. WASHINGTON

A more comprehensive survey was conducted in 2001 in Washington (Johnston *et al*., 2002; see also Johnston *et al*., this issue). Over 1,200 people were surveyed either by questionnaire or oral interview in six coastal communities. A majority of residents (62%) had seen tsunami hazard zone maps and received information on tsunami hazards (76%). However, only 19% of visitors had seen the maps. While they report a high level of general tsunami awareness among residents, they point out that few residents have actually taken actions to prepare for tsunami hazards and most are relying on the actions of governmental agencies and have a limited perception of their own ability to play a role in reducing tsunami hazards.

7.3. NORTHERN CALIFORNIA

Five telephone surveys between 1993 and 2001 were conducted by the Humboldt Earthquake Education Center to assess awareness, preparedness, and the effectiveness of hazard mitigation programs in California's north coast region (Dengler, 2001). Telephone calls were made to randomly selected coastal residents of Humboldt and Del Norte Counties and respondents were asked questions about their awareness of earthquake and tsunami hazards and preparedness actions taken. This study used a panel design and included between 400 and 600 respondents in each survey. The results of the tsunami-related questions are shown in Table I. Over the 9-year period covered by the surveys, the percent knowing what a tsunami is increased from 78 to 98%, what the Cascadia subduction zone is from 16 to 42%.

Three of the surveys included questions to assess the effectiveness of the preparedness magazine *Living on Shaky Ground* (Dengler and Moley, 1995). In all categories, magazine readers had a higher percentage of positive responses; the difference is the most significant regarding awareness of the Cascadia subduction zone, the most likely local tsunami source (Table II).

8. Summary and Discussion

The NTHMP has designated education as a primary tool for reducing the tsunami risk to coastal communities in the United States from near-source tsunamis. The Strategic Implementation Plan (Dengler, 1998), developed in the first year of the program to guide mitigation activities, defines education as the first of five planning areas and outlines a number of education goals. During the first 6 years of the program, all five states developed a variety of education materials either supported by NTHMP funds or through contact with NTHMP projects and personnel. Prior to the program, there were few tsunami education materials outside of Hawaii and almost none that addressed the near-source tsunami hazard. Assessment surveys in three of the states suggest the program has succeeded in increasing awareness of tsunamis. However, these surveys were limited in scope. Each used different instruments and methodologies. Only one study was repeated and addressed changing attitudes over time. The existing assessments cannot be used to compare the effectiveness of different state programs or different educational products. The Washington study (Johnston *et al.*, 2001) suggests that although awareness has increased, the public has taken relatively few actions to reduce their tsunami risk and still views tsunami hazard mitigation as a government activity rather than a personal one. Future mitigation efforts of the NTHMP should include the development and implementation of a uniform assessment tool to test the effectiveness of program education products and projects.

Acknowledgements

The mitigation efforts described are the cumulative efforts of many people at the federal, state, and local levels. I particularly thank Mitigation Subcommittee members Greg Crawford, Mark Darienzo, Rich Eisner, George Priest, Chris Jonientz-Trisler, Scott Simmons, and Brian Yanagi.

References

Atwater, B. F., Cisternas, V. M., Bourgeois, J., Dudley, W. C., Hendley II, J. W., and Stauffer, P. H.: 1999, Surviving a Tsunami – Lessons from Chile, Hawaii, and Japan. USGS Circular 1187, 18 pp.

Atwater, B. F., Cisternas, V. M., Bourgeois, J., Dudley, W. C., Hendley II, J. W., and Stauffer P. H.: 2001, Sobreviviendo a un tsunami: lecciones de Chile, Hawai y Japòn. U.S. Geological Survey Circular 1218, 18 pp.

Bernard, E. N.: 1998, Program aims to reduce impact of tsunamis on Pacific states. *Eos, Trans. AGU* **79**(22), 258, 262–263.

Dengler, L.: 1998, Strategic Implementation Plan for Tsunami Mitigation Projects, approved by the Mitigation Subcommittee of the National Tsunami Hazard Mitigation Program, April 14, 1998. Technical Report NOAA Tech. Memo. ERL PMEL-113 (PB99-115552),

NOAA/Pacific Marine Environmental Laboratory, Seattle, WA. http://www.pmel.noaa.gov/pubs/PDF/deng2030/deng2030.pdf.

Dengler, L.: 2001, Tsunami mitigation efforts on California's north coast. In: *Proceedings of the International Tsunami Symposium 2001 (ITS 2001)* (on CD-ROM), NTHMP Review Session, R-14, Seattle, WA, 7–10 August 2001, pp. 187–203. http://www.pmel.noaa.gov/its2001/.

Dengler, L. and Moley, K.: 1995, *Living On Shaky Ground, How to Survive Earthquakes and Tsunamis on the North Coast*. Humboldt Earthquake Education Center, Humboldt State University, Arcata, CA 95521, 24 pp.

González, F. I. and Bernard, E. N.: 1993, The Cape Mendocino Tsunami. *Earthquakes and Volcanoes* **23**(3), 135–138.

Good, J. W.: 1995, Tsunami Education Planning Workshop, Findings and Recommendations. NOAA Tech. Memo. ERL PMEL-106, 41 pp. (PB95-195970), NOAA/Pacific Marine Environmental Laboratory, Seattle, WA.

Johnston, D., Paton, D., Crawford, G., Ronan, K., Houghton, B., and Buergelt, P.: 2005, Measuring tsunami preparedness in Coastal Washington, United States. *Nat. Hazards*. **35**, 173–184 (this issue).

Johnston, D. M., Driedger, C., Houghton, B., Ronan, K., and Paton, D.: 2001, Children's risk perceptions and preparedness: A hazard education assessment in four communities around Mount Rainier, U.S.A. – Preliminary results. Institute of Geological and Nuclear Sciences Science Report 2001/02.

Johnston, D. M., Paton, D., Houghton, B., Becker, J., and Crumbie, G.: 2002, Results of the August–September 2001 Washington State Tsunami Survey. Institute of Geological and Nuclear Sciences Science Report 2002/17, 35 pp.

Jonientz-Trisler, C. and Mullin, J.: 1999, 1997–1999 Activities of the Tsunami Mitigation Subcommittee: A Report to the Steering Committee NTHMP. FEMA Region 10 publication, 45 pp., Appendices.

Jonientz-Trisler, C., Simmons, R. S., Yanagi, B., Crawford, G., Darienzo, M., Eisner, R., Petty, E., and Priest, G.: 2005, Planning for communities at risk of tsunamis. *Nat. Hazards* **35**, 121–139 (this issue).

Karel, A.: 1998, Oregonians need more information about tsunamis to save lives (results of a survey). *Oregon Geol.* **60**(3).

Oppenheimer, D. H., Beroza, G., Carver, G., Dengler, L. A., Eaton, J. P., Gee, L., González, F., Magee, M., Marshall, G., Murray, M., McPherson, R. C., Ramanowicz, B., Satake, K., Simpson, R., Somerville, P., Stein, R., and Valentine, D.: 1993, The Cape Mendocino, California, earthquake of April 1992: Subduction at the triple junction. *Science* **261**, 433–438.

Public Affairs Working Group (PAWG): 1998, *National Tsunami Hazard Mitigation Program Tsunami Public Affairs Working Group Report Anchorage*. http://www.pmel.noaa.gov/tsunami-hazard/pawgoctober98report.html.

Tsunami Hazard Mitigation Federal/State Working Group: 1996, Tsunami Hazard Mitigation Implementation Plan – A Report to the Senate Appropriations Committee, 22 pp., Appendices, http://www.pmel.noaa.gov/tsunami-hazard/hazard3.pdf.

Washington Military Department, Emergency Management Division: 2001a, *Move To High Ground, Tsunami Curriculum Grades K–6*. http://www.prh.noaa.gov/itic/library/pubs/curriculum/tsunami_curriculum. html, 68 pp.

Washington Military Department, Emergency Management Division: 2001b, *Surviving Great Waves of Destruction*. http://www.prh.noaa.gov/itic/library/pubs/curriculum/tsunami_curriculum.html, 52 pp.

Natural Hazards (2005) 35: 155–162

Planning for Tsunami: Reducing Future Losses Through Mitigation

RICHARD K. EISNER*
CISN and Earthquake Programs, Governor's Office of Emergency Services, 724 Mandana Boulevard, Oakland, CA 94610-2421, USA (Tel: +1-510-465-4887; Fax: +1-510-663-5339; E-mail: Rich_Eisner@oes.ca.gov)

(Received: 5 September 2003; accepted: 1 April 2004)

Abstract. The National Tsunami Hazard Mitigation Program is a multi-faceted approach that encompasses tsunami identification, alert and warning systems and a comprehensive approach to tsunami risk reduction. This paper describes efforts to promote land use planning and development practices that reduce tsunami risk by local elected government and administrative officials. Seven Principles of Tsunami Risk Reduction are presented that range from risk assessment to site planning criteria.

Key words: land use planning, coastal development, tsunami risk mitigation

Abbreviations: FEMA – Federal Emergency Management Agency, IBC – International Building Code, NTHMP – National Tsunami Hazard Mitigation Program, UBC – Uniform Building Code

1. Introduction

Guiding development in areas subject to tsunami inundation poses severe problems for land use planners and regulators. In most coastal areas along the Pacific basin, the tsunami threat is uncertain, and in most areas the probability of occurrence and recurrence intervals are not known. Unlike the earthquake threat, winter storm surge threat or general threat of flood, where recurrence data is used for development decision making, the science of tsunami probability can be summarized as attempting to quantify a very low probability but extremely high impact threat.

Adding to the complex dynamic of tsunami mitigation efforts is the high value local governments place on their coastal environment, where the pressures for coastal access collide with private property rights, pressures for development in proximity to the surf line, and the necessary construction of water oriented recreation facilities in the areas at greatest risk. In most of the

* Regional Administrator, California Governor's Office of Emergency Services and Manager, California Integrated Seismic Network, Earthquake and Tsunami Program

coastal states, it is the local government decision makers (city and county councils) that make land use and development decisions.

Faced with these conflicts, the National Tsunami Hazard Mitigation Program developed a strategy that utilizes techniques already incorporated into the broader and more "robust" efforts of coastal zone management, and planning procedures responsive to storm surge, coastal erosion, and coastal preservation, along with background and educational materials to introduce local officials to the tsunami threat. To implement this strategy, the National Tsunami Hazard Mitigation Program (NTHMP) retained a unique, interdisciplinary team of land use planners, engineers, building, and tsunami experts to combine their knowledge about the tsunami threat with their experience with local and state land use and development practices[1]. Drawing on tsunami modeling, land use regulation, architectural design, and site planning procedures, the team brought together a broad range of tools and approaches to address tsunami risks. This approach is incorporated as a comprehensive but simplified approach to tsunami mitigation in *Designing for Tsunamis: Seven Principles for Planning and Designing for Tsunami Hazards* (National Tsunami Hazard Mitigation Program, 2001; Figure 1)[2].

It is estimated that more than 900,000 people in 489 communities in the states of California, Oregon, Washington, Alaska, and Hawaii live in areas vulnerable to a 50-foot tsunami. Tsunami preparedness needs to address both mitigation and evacuation and response planning by local governments. While the danger from distant tsunamis can be communicated to coastal residents by NOAA's Tsunami Warning Centers, only mitigation and education can protect residents from tsunamis generated near their communities.

NTHMP recognized that planning for tsunamis will not be a high priority for most coastal communities, but by integrating the tsunami threat into other community mitigation and education efforts, the safety of coastal populations can be significantly increased.

Designing for Tsunamis is intended as a guide for local government elected officials and appointed planners, zoning officials, building officials, and those responsible for community development and redevelopment. It focuses mitigation through the use of land use and development policy, building design, and site planning. The guideline is a supplement to other publications developed by the NTHMP and the individual states that address emergency

[1]J. Laurence Mintier, Land Use Planning; L. Thomas Tobin, Coastal Engineering and Mitigation; Robert Olson, Government Mitigation Practices; Bruce Race, FAIA, Building Design and Site Planning; Jeanne Perkins, Mitigation; Daniel Jansenson, Architectural Design; James Russell, Building Codes and Regulation; Robert Wiegel, Coastal Engineering; Mark Legg, Geophysics and Tsunami Generation; Costas Synolakis, Tsunami Modeling and Mechanics.

[2]Designing for Tsunamis is supported by technical "white papers" prepared by the contributing author-experts.

Figure 1. Designing for tsunamis: seven principles for planning and designing for tsunami hazards.

response and evacuation planning[3]. In preparing the guide, the authors reviewed the regulatory context for federal, state, and local planning and developed approaches that were consistent and compatible with local planning authorities. Based on their research, the authors identified the following seven basic principles for reducing a community's risk.

1.1. PRINCIPLE 1: KNOW YOUR COMMUNITY'S TSUNAMI RISK, HAZARD, VULNERABILITY, AND EXPOSURE

The foundation for local government planning and regulation is an objective and scientific assessment of the threat. This chapter outlines the nature of tsunamis, the differences between distant (tele-tsunami) and near-source tsunamis, the physics of tsunami wave propagation, the mechanisms of tsunami-caused damage, including flooding, velocity, and debris impact, and a brief history of damaging tsunamis in the Pacific Ocean.

Nearly 900,000 people are within the inundation of a 50-foot tsunami in the Pacific states. More than 152 communities in California with 590,000 residents are at risk, while 337 communities in Alaska, Hawaii, Oregon, and Washington are endangered. The guide outlines a methodology for

[3]For other NTHMP and state preparedness resources, see: http://www.pmel.noaa.gov/tsunami-hazard/links.htm.

identifying a community's risk, and recommends the use of tsunami specialists for preparing scenarios and loss studies. A tsunami scenario provides the basis for public education of the risk, constituency building for mitigation programs, and response and evacuation planning by emergency managers. An earthquake and tsunami scenario prepared for Humboldt and Del Norte counties in California is included in the guide to illustrate the content and uses of a scenario (California Geological Survey, 1995).

1.2. PRINCIPLE 2: AVOID NEW DEVELOPMENT IN TSUNAMI RUN-UP AREAS TO MINIMIZE FUTURE TSUNAMI LOSSES

A key to long-term reduction of community risk is the use of land use planning processes to guide future development. The objective is to reduce new development at risk so that future losses are minimized. Comprehensive plans, zoning ordinances and subdivision regulations can determine the location, density, types of development, and pattern of development in order to reduce risk. In coastal communities, these tools need to be reviewed to ensure that they address the tsunami risk. Where practical, use of open space setbacks, designation as low density, or recreation use or acquisition reduce potential losses. Subdivision and site planning regulations can be used to guide construction into less vulnerable locations in the inundation zones. While planning and zoning cannot prohibit coastal development, they can ensure that the type and location of permitted development is appropriate to the risk in the inundation zone. Since the opportunities for coastal development and risk mitigation vary with the local political and economic context, there is no one size plan to fit all coastal communities. The planning approach and acceptable level of risk will be determined in each community.

1.3. PRINCIPLE 3: LOCATE AND CONFIGURE NEW DEVELOPMENT THAT OCCURS IN TSUNAMI RUN-UP AREAS TO MINIMIZE FUTURE TSUNAMI LOSSES

The third principle emphasizes project review, site planning, and configuration of buildings to reduce their vulnerability to tsunami damage. The guide proposes the creation of a development review process that emphasizes the incorporation of mitigation techniques at project inception, an approach that ensures that new development incorporates a community's mitigation priorities. It is also considerably easier to reduce risk in the design of new developments than it is to retrofit existing developments that have ignored the tsunami threat. Mitigation approaches should incorporate site geology and topography to locate structures in areas not subject to inundation, design and elevation of buildings above projected flood levels, and attention to structural design to ensure that foundations and structures can withstand expected earthquake and tsunami forces. Site planning should also be used to slow and channel inundation away from structures. There are numerous structural techniques identified in the Federal Emergency Management Agency's (FEMA) Coastal Construction Manual that are applicable to both storm surge and tsunami inundation (FEMA, 2002).

The guide includes a case study of the reconstruction of the city of Hilo, Hawaii, where devastating tsunamis in 1946 and 1960 prompted the formulation of a Downtown Development Plan to ensure that new development and redevelopment reduce future tsunami and flood losses.

1.4. PRINCIPLE 4: DESIGN AND CONSTRUCT NEW BUILDINGS TO MINIMIZE TSUNAMI DAMAGE

Where land use and site planning determine that structures are built in areas subject to tsunami inundation, construction techniques, building materials, enhanced engineering design, and building configuration can help to reduce property damage in future tsunamis.

The guide provides "performance objectives" for buildings in tsunami inundation zones, including location and configuration, elevation, structural and non-structural design standards, structural materials, and location of utilities. Also discussed are approaches for evaluating potential for tsunami damage against the criteria for performance.

Most local communities in the Pacific states have adopted the Uniform Building Code or the International Building Code prepared by the International Conference of Building Officials (ICBO).

Where the UBC or IBC is required, plan review and code enforcement is the responsibility of the local government, resulting in some variation in the quality of construction. While the UBC/IBC address fire, wind, earthquake, and flood design, there are no specific requirements for tsunami resistant construction. This places an additional responsibility on the designer and local code agency to ensure quality construction. The building code from the City and County of Honolulu and FEMA's Coastal Construction Manual provide guidance to architects and engineers in addressing tsunami forces.

Designing for Tsunamis points out that good design must be responsive to a community's needs, and should address codes and standards for a range of hazards, locally validated threat assessments, an objective determination of threat magnitudes, and a determination of a building's performance objectives by the owner, architect, and structural engineer.

1.5. PRINCIPLE 5: PROTECT EXISTING DEVELOPMENT FROM TSUNAMI LOSSES THROUGH REDEVELOPMENT, RETROFIT, AND LAND REUSE PLANS AND PROJECTS

As a community evolves, there are recurring opportunities to build mitigation into development and redevelopment plans. The guide outlines a process that identifies opportunities for gradually improving community safety and resilience through identification of at-risk areas, evaluation of proposals for redevelopment, and the retrofit and reuse of existing structures. Options include the demolition of at-risk structures, providing financial incentives to encourage mitigation, adoption of special code requirements for the retrofit of structures in inundation zones, and the requirement for review of design by specifically qualified architects and structural engineers or peer review committees. The objective of each recommendation is to ensure that attention to the tsunami threat is addressed in planning and community redevelopment decision making.

1.6. PRINCIPLE 6: TAKE SPECIAL PRECAUTIONS IN LOCATING AND DESIGNING INFRASTRUCTURE AND CRITICAL FACILITIES TO MINIMIZE TSUNAMI DAMAGE

Critical infrastructure are those water and power utilities and facilities such as hospitals, fire and police stations that are essential to a community's safety. Loss of such facilities during an earthquake or tsunami event would leave a community helpless to respond, so decisions on their location and construction must be carefully considered. Since many utilities are private or owned by special districts, mitigation programs need to include participation by government, utility operators, and development interests.

The guide recommends the adoption of a comprehensive risk management policy that includes all stakeholders with interests in the coastal inundation zone. The objective of the policy should be the continued operation of critical infrastructure during and after earthquakes and tsunami events. New infrastructure should be designed and located to minimize future disruption. An inventory of existing facilities and risk assessment should provide the basis for on-going mitigation and risk reduction investments that relocate facilities out of harm's way, or strengthen existing facilities to withstand expected tsunami forces.

1.7. PRINCIPLE 7: PLAN FOR EVACUATION

Many existing communities lie within tsunami inundation zones. While decisions can now be made to limit future development at risk, mitigation and redevelopment actions will require decades of planning and investment to reduce the risk in existing communities. Developing local evacuation plans and procedures is therefore essential to protecting coastal residents from tsunami events. The guide discusses horizontal (out of buildings to high ground) and vertical (within buildings to upper floors) evacuation as options, depending on location and structure type, and provides a process for developing a plan and strategy for evacuation. Additional resources are available in *Local Planning Guidance on Tsunami Response* published by the State of California (California Governor's Office of Emergency Services, 1998).

2. The Challenge

Land use and development decisions that will reduce losses from future tsunamis rest with local governments in most states. The challenge for the national and state programs is to provide local decision makers with credible data on the threat of tsunami, and cost effective tools for reducing risk, without the imposition of unreasonable constraints on coastal development.

The National Tsunami Hazard Mitigation Program recognizes that reducing future life loss and property damage can only be achieved by a long-term commitment to risk mitigation to eliminate the threat posed to our communities. Mitigation will take decades to accomplish, so the NTHMP has developed a comprehensive approach that bolsters mitigation achievements with advancements in the understanding of tsunami generation and propagation, improvements in tsunami detection, warning systems, modeling of tsunami effects, and in public education to protect coastal residents and reduce future losses.

The NTHMP also recognizes that reducing the risks posed by tsunami is not the responsibility of a single discipline, or that one solution will solve the problem in every jurisdiction. In some communities, risk reduction will be accomplished through land use planning and development regulation. In other communities, engineering structures may be a necessary "fix." *Planning for Tsunamis* provides a range of tools; the appropriateness of each will be determined at the local government, local political level.

Planning For Tsunamis was intended to be one tool in a package of resources for local government decision makers. This guide has been distributed to land use planning and development agencies in coastal communities subject to tsunami. It was intended to be a companion document to FEMA's Coastal Construction Manual in defining mitigation approaches to reduce risk. Implementation of the guide's risk reducing prescriptions will, however, ultimately depend on the availability of probabilistic risk assessments with the same credibility of flood, landslide, coastal storm inundation, and earthquake faulting; and the creation of a tsunami hazard regulatory framework that does not now exist in most states.

Planning For Tsunamis is the first step in a long-term commitment to tsunami risk reduction by the NTHMP. It has provided awareness of the threat to local decision makers and planners. Future initiatives should include dissemination of the guide through professional associations of planners and government councils, the development of regulatory tools to implement risk reduction, and local workshops and training on implementation approaches.

References

California Geological Survey: 1995, Planning Scenario in Humboldt and Del Norte Counties, California for a Great Earthquake on the Cascadia Subduction Zone. CDMG Pub. 115.

California Governor's Office of Emergency Services: 1998, Local Planning Guidance on Tsunami Response: A Supplement to the Emergency Planning Guidance for Local Governments. State of California, Oakland.

FEMA: 2002, Coastal Construction Manual, Third Edition. FEMA Publication 55 CD, available from the Federal Emergency Management Agency, Department of Homeland Security, at www.fema.gov.

National Tsunami Hazard Mitigation Program: 2001, Designing for Tsunamis: Seven Principles for Planning and Designing for Tsunami Hazards. NOAA, J. Lawrence Mintier *et al.*, Sacramento, CA, 60 pp. Available as a PDF file at: http://www.pmel.noaa.gov/tsunami-hazard/links.html#multi, along with detailed technical background papers.

Natural Hazards (2005) 35: 163–171

NOAA Weather Radio (NWR) – A Coastal Solution to Tsunami Alert and Notification

GEORGE L. CRAWFORD
Washington State Military Department, Emergency Management Division, M/S TA-20, Camp Murray, WA 98430-5211, USA (Tel: +1-253-512-7067; Fax: +1-253-512-7205; E-mail: g.crawford@emd.wa.gov)

(Received: 6 August 2003; accepted: 15 March 2004)

Abstract. The Washington State/Local Tsunami Work Group adopted the NOAA Weather Radio "All-Hazards" Warning System to warn citizens quickly and effectively of not only tsunami hazards but also other natural or man-made hazards. In concert with an array of deep ocean tsunami detectors, land-based seismic sensors, and warning messages issued by the tsunami warning centers, NWR provides a means to expeditiously get critical decision-making information to emergency managers, elected officials, and first responders. To implement the NWR strategy effectively, a partnership was developed to add a repeater to the NWR system to provide complete coverage to the coast of Washington and to shipping lanes off the coast. The Work Group also recognized the need to disseminate time critical hazard information on tsunamis to the public on beaches and in high traffic areas, so it developed a new notification system, with the first prototype installed on 2 July 2003 in Ocean Shores, Washington. A public education program also was developed to improve the impacted communities' understanding of the tsunami hazard, the warning system, and actions they should take if a tsunami occurs.

Key words: NOAA Weather Radio, NOAA Weather Radio "All-Hazard" Warning System, NWR Emergency Information Network, All Hazard Alert Broadcasting Radio, Tsunami Warning Messages

Abbreviations: AHAB Radio – All-Hazard Alert Broadcasting Radio, EAS – Emergency Alert System, EOC – State Emergency Operation Centers, NAWAS – National Warning System, NWWS – NOAA Weather Wire Service

1. Introduction

This paper documents development of a tsunami alert and notification system that is integrated into NOAA Weather Radio's "All-Hazards" Warning System. This system minimizes the potential for erroneous information that may be disseminated to the public through other methods by broadcasting watch and warning information directly from the source. It supplements the communities' communications infrastructure by quickly notifying residents and visitors of the impending tsunami, and warning them to take immediate action to head inland and to higher ground.

2. State/Local Tsunami Work Group

The key ingredient of the Washington State Tsunami Program is the State/Local Tsunami Work Group, composed of representatives from coastal communities, and state and federal agencies. This multi-disciplinary group meets quarterly and invites people from various disciplines to discuss tsunami issues and projects. Using hazard assessment, warning guidance, and mitigation tools developed by the National Tsunami Hazard Mitigation Program, the Work Group developed a mitigation strategy based on needs assessments of individual communities. This process allowed the communities to "buy into" the tsunami program and led to rapid implementation of mitigation and preparedness tools. The Work Group realized that injury and loss of life could be minimized if coastal populations are warned early enough to take appropriate action from an approaching tsunami. In response, local leadership decided to use NOAA Weather Radio as the primary means of alert and notification for communities vulnerable to tsunamis.

3. Warning Guidance Delivery Systems

The National Tsunami Hazard Mitigation Program (NTHMP) works to ensure tsunami warning information is as accurate as possible using real-time data through two systems: deep ocean tsunami detection tsunameters and a NTHMP seismic network. (See papers in this issue by Frank González and David Oppenheimer for more details.) Real-time data provides the West Coast/Alaska and Pacific Tsunami Warning Centers with quick and reliable information to determine whether a seismic event has generated a tsunami. If the event is tsunamigenic, the information is sent to decision-makers. Figure 1 shows the tsunami warning/evacuation communications cycle that quickly provides data to help decision makers understand the scope and complexity of the impending tsunami threat and allow them to make sound decisions to reduce the impact of that threat.

4. NWR Emergency Information Network

The State/Local Tsunami Work Group developed a program to supplement the communities' communications infrastructure to improve local access to emergency information. The Work Group established an "Emergency Information Network" Program that installed NOAA Weather Radio receivers at designated Emergency Information Centers, including visitors centers, hotels and motels, marinas, parks, gas stations, and grocery stores. NWR placards (Figure 2) have been visibly posted at those sites and explanatory brochures distributed through local Chambers of Commerce. Additionally, local emergency managers have encouraged residents to purchase NOAA Weather Radio receivers for their homes, cars, and outdoor activities.

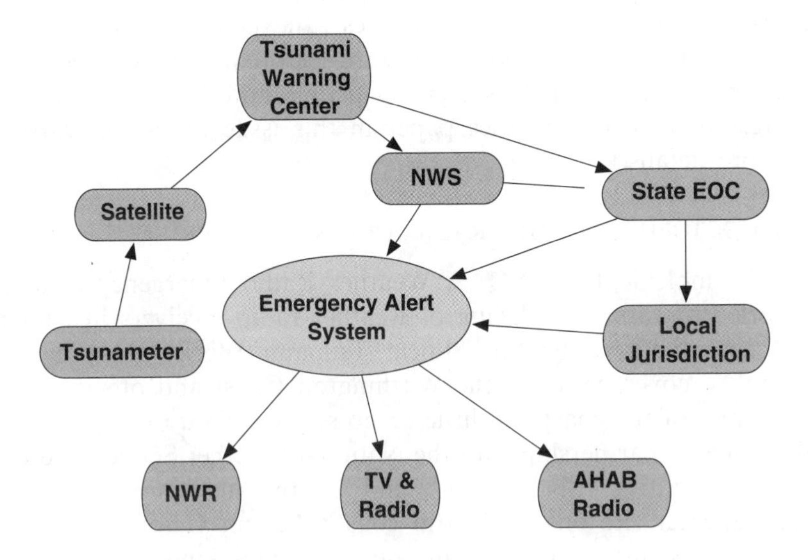

Figure 1. Tsunami Warning/Evacuation Communications Cycle that interfaces with tsunameters, the Tsunami Warning Centers, National Weather Service, and State Communication Systems.

Figure 2. Window sticker that identifies locations with a NOAA Weather Radio receiver.

5. NOAA Weather Radio

The State/Local Tsunami Workgroup supports use of NOAA Weather Radio as an effective and primary all-hazard alert and notification system in tsunami-threatened coastal communities. Inexpensive weather radio receivers can warn listeners about a hazard before the mass media and county alerting systems can do so, giving people additional time to react before danger hits their area. (See paper in this issue by Mark Darienzo *et al.* for more details.)

5.1. MOUNT OCTOPUS WEATHER RADIO TRANSMITTER PROJECT

To effectively implement the NOAA Weather Radio Emergency Information Network Program and placing of weather radio receivers in tsunami threatened communities, the State/Local Tsunami Work Group had to ensure complete coverage along the Washington Coast and offshore shipping lanes. Parts of the coast had little or no service. To do this, the Work Group developed a partnership with the National Weather Service, the U.S. Navy, coastal counties, tribal nations, and the private sector to install a new NWR repeater site on the central coast's Mount Octopus. The new transmitter site became operational in late 2000. This kind of partnership was the first in the nation to establish complete coastal NWR coverage (Figure 3).

6. All Hazard Alert Broadcasting (AHAB) Radio System

NOAA Weather Radio is gaining popularity in the coastal communities. While the state has been successful in deploying weather radio receivers throughout these communities, it lacked a notification system that could be placed on remote beachheads and other highly trafficked areas. The Work Group decided to utilize the capabilities of NWR to quickly disseminate warning messages to those in remote coastal locations. In a brainstorming session, the concept of the AHAB Radio was developed. The first prototype system was installed in Ocean Shores, Washington on 2 July 2003 (Figure 4).

6.1. AHAB RADIO CONCEPT

Coastal communities needed a reliable outdoor alert and notification system that is capable of providing all-hazard warning messages, one that is economical, reliable, and easy to understand and use, and hardy enough to withstand gale-force winds and salt corrosion, require little to no maintenance, and could be placed in areas without electrical power service. Capabilities of the pole-mounted system included (Figures 5 and 6):

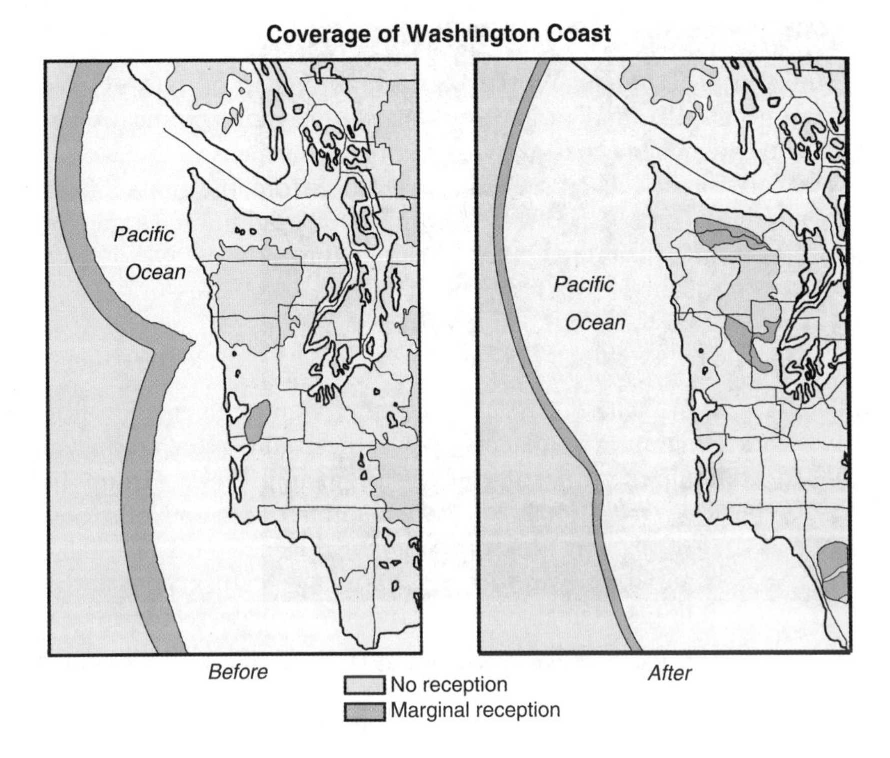

Figure 3. NOAA Weather Radio reception before and after installation of the new NOAA Weather Radio transmitter at Mount Octopus.

- A bright blue strobe light – the same type used by the Coast Guard to cut through fog and be seen from a long distance
- Modulator speaker – with 360° coverage to provide a voice message and coverage to a small community
- Battery operated – charged by wind, and solar, or commercial power
- Triggered by:
  - Hazard Event Code – example: NWS Tsunami Warning Message
  - A Specific Location Code – example: County or city law enforcement and emergency managers

7. Alert and Notification Communications Flow

When data indicates a tsunami has been generated, the Tsunami Warning Centers transmit the appropriate message (Information statement, Advisory, Watch, or Warning) on the NOAA Weather Wire Service (NWWS). The National Weather Service offices, newswires, and state teletype system receive the message. The Tsunami Warning Centers also transmit an oral notification

Figure 4. AHAB Radio prototype installed at Ocean Shores, Washington on 2 July 2003.

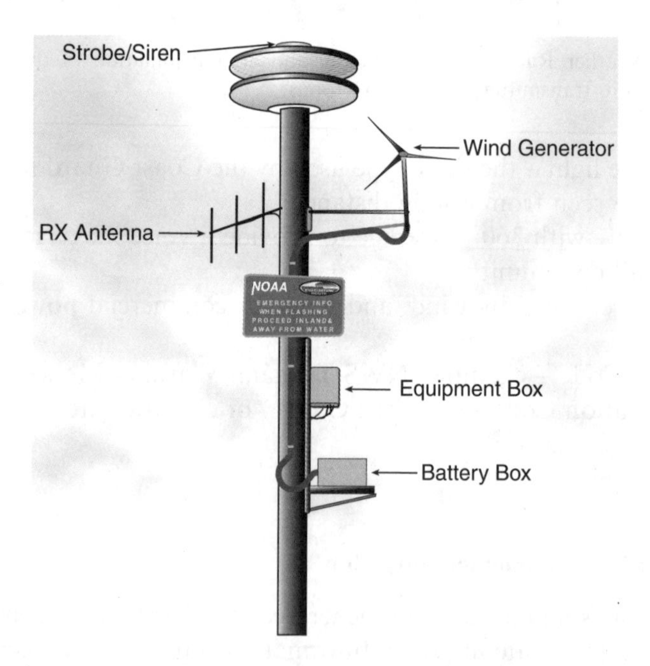

Figure 5. Conceptual drawing of AHAB Radio developed by George Crawford and Michael Namchek, Washington Emergency Management Division in partnership with Federal Signal.

36" x 36"
3"C LETTERS
BLUE - BACKGROUND (REFLECTIVE)
WHITE - LEGEND (REFLECTIVE)

Figure 6. Information sign placed on the pole of the Tsunami Notification System. The sign provides information on what to do when the bright blue light at the top of the pole flashes.

to State Emergency Operation Centers (EOC) via the National Warning System (NAWAS). The State EOC then passes the message to 24-hour contact points in affected jurisdictions via NAWAS and the state teletype system. The local jurisdiction's 24-hour contact points then notify emergency management personnel, key responders, and elected officials. If a tsunami warning message is received and evacuation is ordered by the local jurisdiction, an Emergency Alert System (EAS) message is transmitted over the local EAS Relay Network by local authorities. At the request of local authorities, the State EOC also can transmit the EAS message. The EAS message is automatically relayed over coastal NWR transmitters, reaching AHAB Radio, commercial broadcast stations (AM, FM, TV, and cable), and all of those with weather radio receivers programmed to receive the EAS message (Figure 7).

8. Education

The State/Local Tsunami Work Group has developed tsunami brochures to provide information on the tsunami hazard. These brochures include

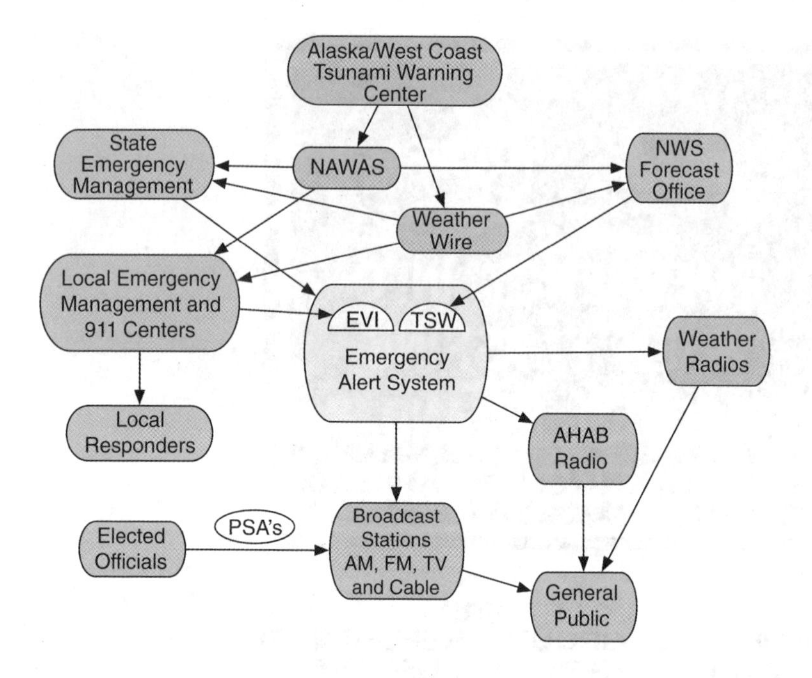

Figure 7. Identifies the communications flow of a tsunami warning message from the Alaska/West Coast Tsunami Warning Center to the general public. It also identifies the routing of EAS messages.

evacuation maps, NOAA Weather Radio information, and tsunami safety tips. The State worked with David Johnston from the Institute of Geological and Nuclear Sciences in New Zealand to examine residents' and visitors' perception of the tsunami hazard. The study also looked at their knowledge of the Washington Warning System and their understanding of what to do if a tsunami were to strike. (See paper in this edition by David Johnston for more details.) Results from Johnston's study were used to revise the booklet "How the Smart Family Survived a Tsunami" (elementary edition—K–6). The booklet now addresses the tsunami warning process, AHAB Radio, and actions people should take when a tsunami warning is received. It also has information on a family disaster plan and disaster supply kit.

With September designated as Weather Radio Awareness Month in Washington, the Work Group's goal is to have NOAA Weather Radio receivers become as common as smoke detectors in homes and businesses statewide to help protect lives and property from natural and technological hazards.

Acknowledgements

The successful development of these products and the review of this paper could not have been possible without the support of the Mitigation and

Telecommunications Sections, Emergency Management Division. Special thanks to Michael Namchek, Telecommunications Coordinator, Emergency Management Division, who, with the author, developed the AHAB Radio from concept to implementation. Also, Ted Buehner, Warning Coordination Meteorologist, National Weather Service, Seattle has been a major contributor in the development of these products and has been a key partner in the development of the annual Weather Radio Awareness Month campaign in Washington.

References

Crawford, G. L.: 2001, Tsunami inundation preparedness in coastal communities. In: *Proceedings of the International Tsunami Symposium 2001 (ITS 2001)* (on CD-ROM), NTHMP Review Session, R-18, Seattle, WA, pp. 214–218. http://www.pmel.noaa.gov/its2001

Johnston, D., Paton, D., Houghton, B., Becker, J., and Crumbie, G.: 2002, *Results of the August–September 2001 Washington State Tsunami Survey*. Institute of Geological and Nuclear Sciences Limited, Lower Hutt, New Zealand.

Oregon Emergency Management and the Oregon Department of Geology and Mineral Industries: 2001, Tsunami Warning Systems and Procedures Guidance for Local Officials. ODGAMI Special Paper 35.

Natural Hazards (2005) 35: 173–184

Measuring Tsunami Preparedness in Coastal Washington, United States

D. JOHNSTON[1★], D. PATON[2], G. L. CRAWFORD[3], K. RONAN[4], B. HOUGHTON[5] and P. BÜRGELT[4]
[1]*Institute of Geological and Nuclear Science, 69 Gracefield Road, PO Box 30-368, Lower Hutt, New Zealand;* [2]*University of Tasmania, Launceston, TAS 7250, Australia;* [3]*Washington State Military Department, Emergency Management Division, M/S TA-20 Camp Murray, WA 98430-5211, USA;* [4]*Massey University, Palmerston North, New Zealand;* [5]*University of Hawaii at Manoa, 2500 Campus Road, Honolulu, HI 96822, USA*

(Received: 26 September 2003; accepted: 7 April 2004)

Abstract. A survey of over 300 residents' and visitors' (non-residents) perceptions of tsunami hazards was carried out along the west coast of Washington State during August and September 2001. The study quantified respondents' preparedness to deal with tsunami hazards. Despite success in disseminating hazard information, levels of preparedness were recorded at low to moderate levels. This finding is discussed in regard to the way in which people interpret hazard information and its implications for the process of adjustment adoption or preparedness. These data are also used to define strategies for enhancing preparedness. Strategies involve maintaining and enhancing hazard knowledge and risk perception, promoting the development of preparatory intentions, and facilitating the conversion of these intentions into sustained preparedness. A second phase of work began in February 2003, consisting of a series of focus groups which examined beliefs regarding preparedness and warnings, and a school survey. Preliminary findings of this work are presented.

Key words: tsunami, public education, warnings, warning systems, preparedness, evacuation

1. Introduction

Considerable improvement in the understanding of tsunami risk in Washington has emerged from research over the past two decades (Wilson and Torum, 1972; Atwater, 1992; Atwater *et al.*, 1995; Walsh *et al.*, 2000). Since the mid-1990s the State of Washington, in association with the U.S. National Tsunami Mitigation Program, has undertaken a wide range of mitigation activities (Jonientz-Trisler and Mullin, 1999; Bernard, 2001). Consequently, information in several media (books, posters, pamphlets, school kits, mugs, and magnets) has been distributed to the communities surveyed here (Figure 1). Warning and evacuation signs have been erected in prominent positions, and maps and public displays illustrating the tsunami inundation zone

★ Author for correspondence: Tel: +64-4-570 4538; E-mail: d.johnston@gns.cri.nz

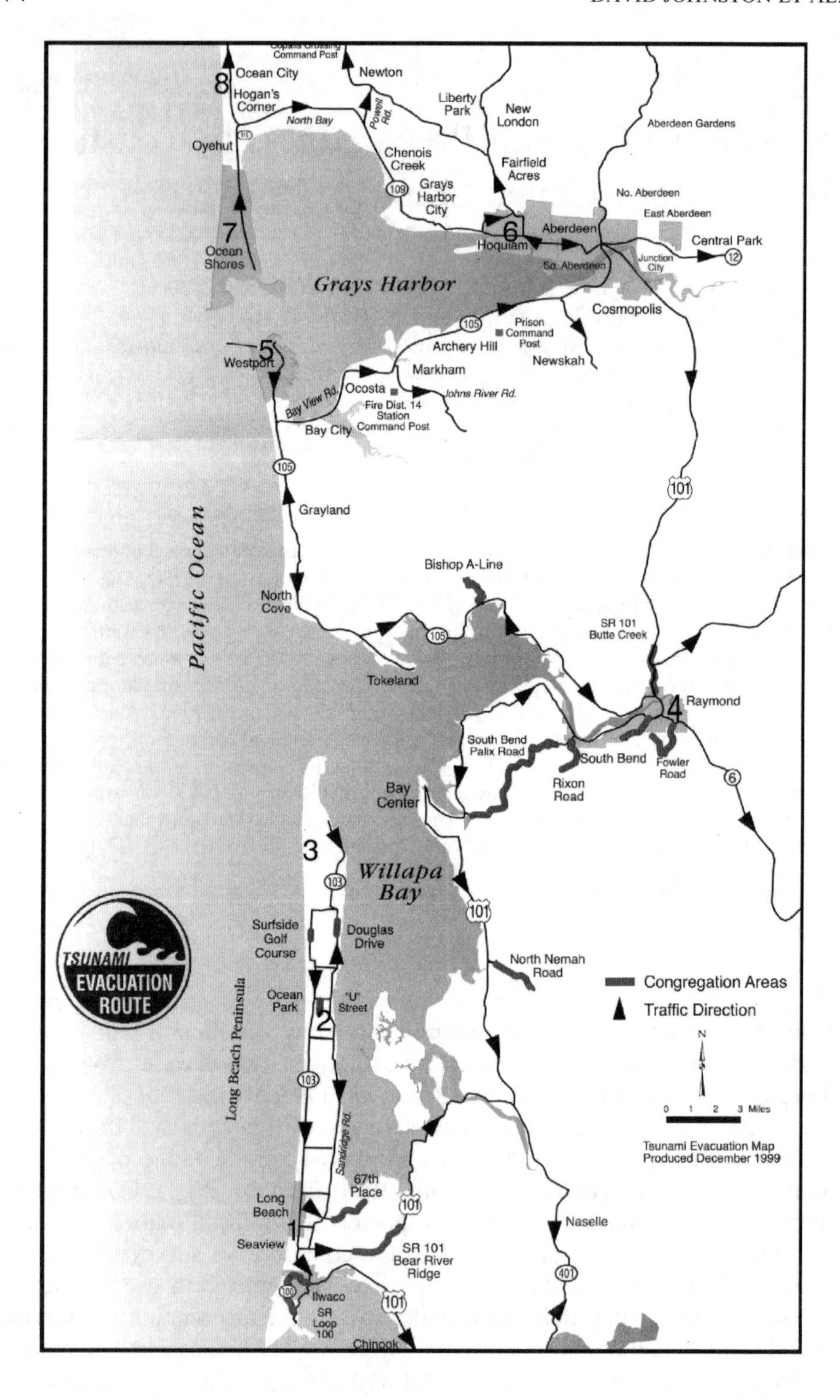
Command Post
Newton
8
Ocean City
Hogan's Corner
North Bay
Powell Rd.
Liberty Park
New London
Aberdeen Gardens
Oyehut
Chenois Creek
Fairfield Acres
Grays Harbor City
No. Aberdeen
East Aberdeen
7
Ocean Shores
6
Aberdeen
Hoquiam
Central Park
Junction City
So. Aberdeen
Grays Harbor
Cosmopolis
Prison Command Post
Archery Hill
5
Westport
Newskah
Markham
Bay View Rd.
Ocosta
Johns River Rd.
Fire Dist. 14 Station Command Post
Bay City
Grayland
Pacific Ocean
Bishop A-Line
North Cove
SR 101 Butte Creek
Tokeland
4
Raymond
South Bend Palix Road
South Bend
Fowler Road
Rixon Road
Bay Center
3
Willapa Bay
Surfside Golf Course
Douglas Drive
TSUNAMI
EVACUATION ROUTE
North Nemah Road
Congregation Areas
Traffic Direction
Ocean Park
"U" Street
2
Long Beach Peninsula
Sandridge Rd.
0 1 2 3 Miles
Tsunami Evacuation Map
Produced December 1999
67th Place
Long Beach
1
Seaview
Naselle
SR 101 Bear River Ridge
Ilwaco
SR Loop 100
Chinook

for the southern Washington coast have been distributed to the community. Three studies have recently been undertaken to assess the influence of these activities on tsunami hazard preparedness.

2. Survey

A survey of over 300 residents' ($n = 217$) and visitors' (non-residents, $n = 83$) perceptions of tsunami hazards was carried out along the west coast of Washington State during August and September 2001 (Figure 1). Three different methods were used to collect information: delivering written questionnaires to individual residential houses, using postal (P.O. Box) delivery for questionnaires, and person-to-person interviews with tourists and residents. A total of 436 questionnaires were delivered directly to houses in the communities of Long Beach, Seaview, Ocean Park, Surfside Estates, Oysterville, and Ocean Shores between 26 August and 1 September 2001. A further 733 postal questionnaires were sent to random post office box numbers in the communities of Raymond, Hoquiam, Ocean Shores, and Westport in September 2001. Return rates varied from around 24% in Long Beach/Seaview to 9% in Raymond and provide a moderately representative sample of residents from the area being surveyed. It is also interesting to speculate on the implications of the differential rates of return from each area. Rates of return appear, with a few exceptions, to mirror proximity to the ocean, and thus the source of the tsunami hazard. For example, returns are relatively high from areas directly fronting the ocean: Long Beach (24%), Ocean Park (22%), Ocean Shores (20%), and Westport, (18%).

A total of 97 interviews were also conducted at several West Coast beaches including Long Beach, Seaview, Ocean City, Ocean Shores, and Westport between 28 August and 30 August 2001. People interviewed were mostly visitors (83) but a small number of residents (14) were also included in the sample.

The study was concerned with quantifying people's understanding of tsunami hazards on the Washington coast, their knowledge regarding the Washington State tsunami warning system, their preparedness to deal with tsunami activity, and providing information that could be used for baseline measurement. Data were collected using a questionnaire derived from a theoretically robust and empirically tested process model of preparedness (Paton, 2000, 2003; Paton *et al.*, 2001, 2003b). Details of the scales used and their sources are listed in Table I. A detailed report on the findings is

Figure 1. Survey locations: (1) Long Beach/Seaview, (2) Ocean Park, (3) Surfside/Oysterville, (4) Raymond, (5) Westport, (6) Hoquiam, (7) Ocean Shores, (8) Ocean City. Map also shows planned evacuation routes as presented in Grays Harbor and Pacific Counties' tsunami hazard brochure.

Table I. Scales used and their sources

Scale	Source
Precursor variables	
Risk perception	Johnston *et al.* (1999)
Critical awareness	Dalton *et al.* (2001)
Intention formation variables	
Outcome expectancy	Bennett and Murphy (1997)
Self-efficacy	Paton *et al.* (2001)
Intentions	
Intention/information search	Bennett and Murphy (1997)
Moderator variables	
Response efficacy	Lindell and Whitney (2000)
Perceived responsibility	Mulilis and Duval (1995)
Sense of community	Paton *et al.* (2001)
Timing	Paton *et al.* (2003b)
Outcome	
Adjustment adoption	Mulilis-Lippa Preparedness Scale Mulilis *et al.* (1990)

presented by Johnston *et al.* (2002) and key issues emerging from the study are discussed further by Paton *et al.* (2003a). The interviews with visitors consisted of eight brief questions that asked about the respondents' knowledge of tsunami hazards in the area and their awareness of the Washington State tsunami warning system.

Current initiatives appeared to be moderately to highly effective in raising public awareness of the hazard. For example, 62% of residents had seen the tsunami hazard zone maps and 76% of residents had heard or received information on tsunami hazards from a range of sources. In addition some 68% of residents reported that they had heard or observed other people preparing for tsunami hazards. However, visitors (non-residents) surveyed were significantly less aware of the tsunami hazard and the warning system. For example, only 19% of visitors had seen the tsunami hazard zone maps and 46% were unaware of the elements of the tsunami warning system. These observations suggest a need for additional research on tourist perceptions of, and responsiveness to, warnings and to investigate local attitudes to the provision of this information.

In addition to enhancing hazard knowledge, a second objective of public education programs is to facilitate preparedness to deal with hazard consequences. That is, the degree to which knowledge and awareness translate into preparedness behavior. An examination of the number of preparedness items adopted (Table II) suggests that receipt of the hazard and preparedness information did not translate into a corresponding level of preparedness. Of the 11 adjustments, the average number adopted per household was 2.66 and

Table II. Hazard preparedness indicators and the proportion of residents adopting each

Protect breakable household items	19%
Put strong latches on cabinet doors	7%
Add edges to shelves to keep things from sliding off	5%
Strap water heater	23%
Install flexible tubing to gas appliances	12%
Bolt house to foundation	31%
Pick an emergency contact person outside of the Northwest	28%
Buy additional insurance (e.g., home)	33%
Find out if you are in an area particularly vulnerable to a disaster	57%
Have home inspected for preparedness	3%
Talked to family members about what to do if a tsunami warning is heard	48%

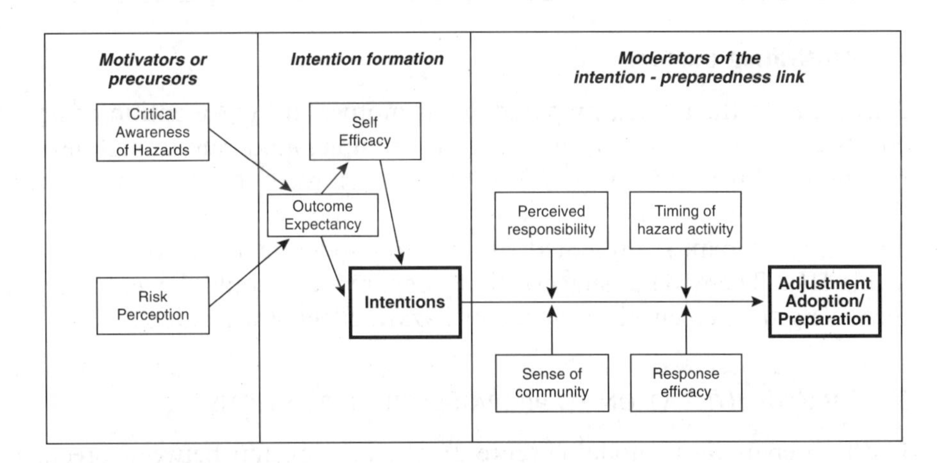

Figure 2. The social-cognitive preparation model. Adapted from Paton *et al.* (2003b).

levels of adoption of each measure were low (Table II). Explanations for this discrepancy focus on the interpretive processes that influence how hazard information is rendered meaningful by its recipients. A tendency to overestimate perceived preparedness by extrapolating from the low levels of loss and damage associated with prior hazard experiences to a capability to deal with future events was supported by the data. A propensity to attribute the need for hazard information and preparedness to other members of their community rather than themselves was also evident (Paton *et al.*, 2003a). An explanation for low preparedness has previously been discussed using a process model of preparedness (Paton, 2003; Paton *et al.*, 2003b) that comprised three distinct, but related, stages (Figure 2). Acknowledging the distinction between these stages is important. They comprise different variables and require different intervention strategies to achieve change.

Table III. Means and standard deviations of preparedness process variables

Variable	Scale Min.–Max.	Mean	SD
Risk perception	2–10	7.31	2.47
Critical awareness	2–10	5.09	2.09
Outcome expectancy	2–10	6.81	2.03
Self efficacy	4–20	10.93	2.37
Intention/information search	3–9	4.55	1.66
Responsibility	1–5	4.27	1.03
Response efficacy	5–25	12.36	4.49
Sense of community	9–45	27.73	4.02
Preparation	0–11	2.66	2.05

3. Motivating Factors

According to the model, preparedness is motivated by perception of hazard effects capable of posing a threat. In this sample, moderate to high levels of perceived threat (mean = 7.31) were attributed to tsunami hazards (Table III). The second motivating factors, critical awareness (thinking and discussing tsunami), presented at low to moderate levels (mean = 5.09, Table III). These data suggest that preparedness could be enhanced by increasing the perceived relevance of hazard effects for residents.

3.1. FROM MOTIVATION TO PREPARATORY INTENTIONS

In the preparedness model (Figure 2), the relationship between precursors and intentions is mediated by outcome expectancy and self efficacy. Moderate levels of outcome expectancy (belief that hazard effects can be mitigated by individual efforts) were recorded (mean = 6.81, Table III) and these act to reduce preparedness. Low-moderate levels (mean = 10.93; Table III) of self-efficacy (judgment regarding their capabilities to mitigate hazard effects) will constrain preparedness. Low levels of these variables is consistent with the finding of low to moderate levels of preparedness intentions. Only 13% of the sample indicated a definite intention to actively prepare. These data are consistent with the low to moderate levels of preparation described above (Table II).

3.2. MODERATING THE INTENTION-PREPARATION LINK

The model describes how preparedness can be moderated by several factors. Moderate to high levels of personal responsibility, resource availability (resource efficacy) and sense of community lessen the likelihood of their acting to moderate preparedness. A final moderator is the time frame within which people anticipate the occurrence of the next tsunami (Paton, 2003). Those who anticipated this occurring within the next 12 months were likely to convert their intentions into actual preparedness. In the present sample, only 2% of the sample thought that a tsunami was likely within the next year. Consequently, this variable could significantly moderate the intention-preparedness link.

To facilitate motivation, public education and empowerment strategies are needed (Paton, 2000) that emphasize the salience of hazard issues for community members. Improved preparedness could also accrue from enhancing community members' beliefs in the feasibility of mitigating hazard effects through personal actions (e.g., counter beliefs that hazards have totally catastrophic effects) and enhancing beliefs in personal competency to implement these activities. Changing these factors requires a mix of public education, social policy, training, and empowerment strategies. The third stage, converting intentions into actual behavior, could be enhanced by focusing on encouraging acceptance of a "sooner rather than later" message. It is also important to understand the belief and attitudes that underpin the above responses. To examine this further a series of focus groups were contacted. The focus group discussions also explored members' perceptions of, and beliefs about, tsunami warnings.

4. Focus Groups

In February 2003 a series of six focus groups were run with the aim of exploring residents' experiences and perceptions of tsunami risk and preparedness. Understanding these attitudes is important and requires using qualitative research as a mode of inquiry. The groups were run in Ocean Shores (hotel managers and seniors), Pacific Beach (volunteer fire-fighters), Ocean Park (seniors), Long Beach (Kiwanis), and Aberdeen (Coastal Community Action Program members). Groups were selected to ensure that the views of a diverse and representative range of constituencies were canvassed. All focus groups were taped and were transcribed.

Initial analysis of the content identified a number of key issues. There was a high level of interest and support for participating in the focus groups. Most people expressed appreciation for the opportunity to "have their say" in an open forum. They were also happy for it to be recorded and pleased that "what was said" was going be used in a constructive way. A wide range of topics were covered in the discussions, including risk perceptions, community awareness, issues relating to preparation, response, warnings and evacuation, and mitigation options (discussion included both comment on current initiatives and suggestions on possible options for the future).

In Ocean Shores many of the hotels had been proactive in promoting awareness and preparedness, including staff and customer awareness, staff training, and other mitigation measures. There was some discussion that this may not be the case in all communities with some remaining resistance from the business community in other parts of the state. A wide discussion was had on the potential effectiveness of warning systems. Some concerns regarding the level of community understanding of the warning system and its limitations were expressed. There seems to be some misunderstanding of the likely warning time that may be given. The issue of evacuation was explored in all groups. There is a clear understanding of the need to evacuate but many felt that the road networks would be unable to cope, especially during peak summer and holiday times. Many residents believed it would not be worth attempting to evacuate by car due to the perceived congestion following an evacuation order. It was commonly suggested that it may be better to move to local high points or as far inland as possible within the local area. Many suggestions for improving education among the population were made. School programs were seen as an important way of improving awareness in the community.

The unstructured data collected in the focus groups were systematically analyzed using various grounded theory analysis strategies and the qualitative data analysis programme ATLAS.ti. The researchers followed closely the procedures for open, axial, and selective coding (Strauss and Corbin, 1990; Browne and Sullivan, 1999; Chamberlain, 1999). Throughout the coding, they constantly drew comparisons among and between incidents, text segments, concepts, codes, and focus groups; asked questions of the data; wrote memos; formulated hypotheses; and created networks. These analysis strategies were used to develop a theory. This approach helped maintain balance between creativity, rigor, persistence, and theoretical sensitivity; assisted with grounding explanations in the data; and facilitated identifying links among concepts (Strauss and Corbin, 1990). The outcome was the best achievable fit between the data and their interpretation (Browne and Sullivan, 1999) and a "conceptually rich understanding and systematic integration of low-level descriptions into a coherent account" (Henwood and Pidgeon, 1992, p. 103).

The qualitative analysis of the focus group discussions yielded the following preliminary findings. Difficulties in regard to information distribution and the adequacy of its formatting reduced the capacity of residents to understand the nature of tsunami hazards, resulting in a substantial lack of information regarding tsunami preparation and warnings. The lack of the continuous availability data was identified as problematic, as was the perception that city councils and real estate agents are holding back information from new residents due to fear of negative impact on economic and business activity. Participants also felt that councils held back information for fear of criticism. Finally, the information that was disseminated was perceived as being too general and in a format that residents had difficulties relating to. Information thus needs to be tailored more specifically to cater to the diverse needs and expectations of different groups within the community.

Inadequacies in regard to the content and frequency of dissemination of information reduced residents' knowledge regarding the nature and effects of tsunami, what they could do to prepare personally, and what their communities have in place for responding when a tsunami should occur. Inadequate knowledge, in combination with the highly complex nature of tsunami (i.e., the effect of tsunami depends on so many different factors and their interaction), contributes to the generation and maintenance of misconceptions and uncertainty among residents, increases the likelihood of residents either exaggerating or downplaying the risk and the consequences of tsunamis, and generated many questions in the focus groups. Participants reported that these issues contributed to many residents becoming apathetic. Inadequate knowledge, misconceptions, high uncertainty and/or apathy led, in several ways, to low levels of individual preparedness and high levels of refusals to evacuate. Although residents know about the many things they could do to prepare personally, they tend not to implement them. Furthermore, in the case of emergency kits, even if they had prepared them, many use their contents after a short while and do not renew them regularly.

The participants perceived that the current level of preparedness for natural hazards within their communities is poorer than it was in the 60s and 70s. In addition to the reasons outlined above, limited preparedness was also attributed to a combination of lack of money, the fear of negative effects on the economy, and perceiving the risk of a tsunami as relatively low.

Low levels of personal and community preparedness generated many concerns regarding warnings. In particular, participants were concerned about being able to get out in time for several reasons. First, they did not believe that the warnings would be early enough and/or loud enough or that there are sufficient sirens to cover the area effectively. The former relates to both the speed with which a warning can be issued, and the belief that a seismic event close inshore would reduce the effectiveness of a warning. Secondly, although noticed by many residents, evacuation signs are not known very well, were not specific enough, did not make sense to many residents, and were misleading. For example, some residents reported that following the signs could take you round in circles and that they did not direct one to safe areas. Further, residents were concerned that, with so many people following the signs simultaneously, the roads, and therefore the evacuation routes, will be blocked. Finally, many participants were highly concerned that there is often only one road out of town and that not all four lanes of this road will be available as exit routes due to accidents and people coming into the area (e.g., parents attempting to retrieve their children from school, outside workers). These concerns are especially pronounced for residents living in flat areas and those residing furthest down the roads out of town. As a consequence of these beliefs many residents believe that they would not get out anyway and, therefore, would not self-evacuate when a tsunami occurs.

Overall, the analysis identified that the residents have to negotiate a highly complex decision-making process to figure out whether to respond, and how to respond, to a warning. However, a combination of their inadequate knowledge and the fact that the effect of tsunami depends on so many different factors, resulted in participants being highly unsure with regard to how to make these decisions, particularly within the short time frame available within which to make these decisions. Whether the participants respond and how they respond is influenced by several attitudes and beliefs that must be accommodated in public education programs if the effectiveness of the latter is to be enhanced.

If the effectiveness of the warning system is to be enhanced and evacuation and preparedness encouraged, it is important to acknowledge the reality of these beliefs. Consultation with the community (Paton, 2000) is required to reconcile these beliefs with the goals of the emergency management community and to promote sustained preparedness and readiness within communities vulnerable to tsunami hazards.

5. School Survey

In addition to the focus groups a series of six school surveys were also undertaken in February 2003. The school research builds on a number of studies undertaken over the past several years in Washington, Hawaii, and New Zealand to assess students' understanding and response to natural hazards (Johnston and Houghton, 1995; Johnston and Benton, 1998; Ronan and Johnston 2001, 2003; Gregg *et al.* 2004). Another primary purpose for undertaking such surveys is to establish and strengthen the link between school education programs and home-based preparedness (e.g., Ronan and Johnston, 2001). The questionnaire used in the current series of surveys is based on one developed for the 2000 Mount Rainier study (Johnston *et al.*, 2001; see also Ronan and Johnston, 2001). To date only a preliminary analysis of the survey has been undertaken. Students have a good awareness of the tsunami risk and perceive it to be a possible event within their lifetime. Most students report being involved in education programs and there is evidence that they have interacted with their parents on hazard issues. Some desirable levels of household preparedness appear to exist. Further analysis of these results will be undertaken over the coming year.

6. Conclusion

The overall conclusion of the three studies is that the hazard education program to date has been successful in terms of promoting awareness of and access to information about tsunami hazard among coastal Washington residents. Despite success in disseminating hazard information, levels of preparedness were recorded at low to moderate levels. The findings in these studies emphasized both the importance of accommodating pre-existing beliefs and interpretive processes, and the need for additional strategies to augment existing programs with initiatives that manage these beliefs and perceptions in ways that facilitate preparedness. The use of multiple methods – surveys, focus groups, and school surveys – is designed to enhance the validity of the findings. The data furnished by these analyses also provide baseline data against which subsequent intervention activities can be assessed.

References

Atwater, B. F.: 1992, Geologic evidence for earthquakes during the past 2000 years along the Copalis River, southern coastal Washington. *J. Geophys. Res.* **97**(B2), 1901–1919.

Atwater, B. F., Nelson, A. R., Clague, J. J., Carver, G. A., Yamaguchi, D. K., Bobrowsky, P. T., Bourgeois, J., Palmer, S. P., *et al.*: 1995, Summary of coastal geologic evidence for past great earthquakes at the Cascadia subduction zone. *Earthquake Spectra* **11**, 1–18.

Bennett, P. and Murphy, S.: 1997, *Psychology and Health Promotion*. Open University Press, Buckingham.

Bernard, E. N.: 2001, The U.S. National Tsunami Hazard Mitigation Program Summary. In: *Proceedings of the International Tsunami Symposium 2001* (*ITS 2001*) (on CD-ROM), NTHMP Review Session, R-1, Seattle, WA, 7–10 August 2001, pp. 21–27. http://www.pmel.noaa.gov/its2001/.

Browne, J. and Sullivan, G.: 1999, Analysing in-depth interview data using grounded theory. In: V. Minichello, G. Sullivan, K. Greenwood and R. Axford (eds.), *Handbook for Research Methods in the Health Sciences*. Addison-Wesley, Sydney, pp. 575–611.

Chamberlain, K.: 1999, Using grounded theory in health psychology. In: M. Murray and K. Chamberlain (eds.), *Qualitative Health Psychology: Theories and Methods*. Sage Publications, London, pp. 183–201.

Dalton, J. H., Elias, M. J., and Wandersman, A.: 2001, Community Psychology. Wadsworth, Belmont, CA.

Gregg, C. E., Houghton, B. F., Johnston, D. M., Paton, D., and Swanson, D.: 2004, The perception of volcanic risk in Kona communities from Mauna Loa and Hualalai volcanoes, Hawaii. *J. Volcanol. Geotherm. Res.* **130**(3–4), 179–196.

Henwood, K. L. and Pidgeon, N. F.: 1992, Qualitative research and psychological theorizing. *J. Psychol.* **83**, 97–111.

Johnston, D. M., Bebbington, M. S., Lai, C.-D., Houghton, B., and Paton, D.: 1999, Volcanic hazard perceptions: Comparative shifts in knowledge and risk. *Disaster Prev. Manage.* **8**, 118–126.

Johnston, D. M. and Benton, K.: 1998, Volcanic hazard perceptions in Inglewood, New Zealand. *Aust. J. Disaster Trauma Stud.* **2**, 1–8.

Johnston, D. M., Driedger, C., Houghton, B., Ronan, K., and Paton, D.: 2001, Children's risk perceptions and preparedness: A hazard education assessment in four communities around

Mount Rainier, U.S.A. – Preliminary results. Institute of Geological and Nuclear Sciences Science Report 2001/02.

Johnston, D. M. and Houghton, B.F.: 1995, Secondary school children's perceptions of natural hazards in the central North Island. *New Zealand J. Geogr.* **99**, 18–26.

Johnston, D. M., Paton, D., Houghton, B., Becker, J., and Crumbie, G.,: 2002, Results of the August–September 2001 Washington State Tsunami Survey. Institute of Geological and Nuclear Sciences Science Report 2002/17, 35 pp.

Jonientz-Trisler, C. and Mullin, J.: 1999, 1997–1999 Activities of the Tsunami Mitigation Subcommittee: A Report to the Steering Committee NTHMP. FEMA Region 10 publication, 45 pp., Appendices.

Lindell, M. K. and Whitney, D. J.: 2000, Correlates of household seismic hazard adjustment. *Risk Anal.* **20**, 13–25.

Mulilis, J.-P., Duval, T., and Lippa, R.: 1990, The effects of a large, destructive local earthquake on earthquake preparedness as assessed by an earthquake preparedness scale. *Nat. Hazards* **3**, 357–371.

Mulilis, J.-P. and Duval, T. S.: 1995, Negative threat appeals and earthquake preparedness: A person-relative-to-event (PrE) model of coping with threat. *J. Appl. Soc. Psychol.* **25**, 1319–1339.

Paton, D.: 2000, Emergency Planning: Integrating community development, community resilience and hazard mitigation. *J. Am. Soc. Prof. Emerg. Manage.* **7**, 109–118.

Paton, D.: 2003, Disaster preparedness: a social-cognitive perspective. *Disaster Prev. Manage.* **12**, 210–216.

Paton, D., Millar, M., and Johnston, D.: 2001, Community resilience to volcanic hazard consequences. *Nat. Hazards* **24**, 157–169.

Paton, D., Smith, L., Johnston, D., Johnston, M., and Ronan, K.: 2003a, Developing a model to predict the adoption of natural hazard risk reduction and preparatory adjustments. EQC Project Report, No. 01–479, Wellington, Earthquake Commission, New Zealand.

Paton, D., Smith, L. M., and Johnston, D.: 2003b, When good intentions turn bad: Promoting disaster preparedness. In: *Proceedings of the 2003 Australian Disaster Conference*, Mt. Macedon, Victoria.

Ronan, K. R. and Johnston, D. M.: 2001, Correlates of hazard education programs for youth. *Risk Anal.* **21**, 1055–1063.

Ronan, K. R. and Johnston, D. M.: 2003, Hazards education for youth: A quasi-experimental investigation. *Risk Anal.* **23**(5), 1009–1020.

Strauss, A. and Corbin, J.: 1990, *Basics of Qualitative Research: Grounded Theory Procedures and Techniques.* Sage, London.

Walsh, J., Caruthers, C. G., Heinitz, A. C., Myers, E. P., Baptista, A. C., Erdakos, G. B., and Kamphaus, R. A.: 2000, Tsunami Hazard Map of the Southern Washington Coast: Modeled Tsunami Inundation from a Cascadia Subduction Zone Earthquake. Washington Division of Geology and Earth Resources Geological Map GM-9.

Wilson, B. W. and Torum, A.: 1972, Runup heights of the major tsunami on North American coasts. In: *National Research Council Committee on the Alaska Earthquake, 1972, The great Alaska Earthquake of 1964 – Oceanography and Coastal Engineering*, pp. 3158–3180.